Accelerated Path to Cures

Josep Bassaganya-Riera

Accelerated Path to Cures

 Springer

Editor
Josep Bassaganya-Riera
Nutritional Immunology and Molecular Medicine Laboratory
Biocomplexity Institute of Virginia Tech
Blacksburg, VA
USA

Biotherapeutics Inc.
Blacksburg, VA
USA

ISBN 978-3-030-10339-2   ISBN 978-3-319-73238-1   (eBook)
https://doi.org/10.1007/978-3-319-73238-1

Printed on acid-free paper

This Springer imprint is published by the registered company Springer International Publishing AG part of Springer Nature.
The registered company address is: Gewerbestrasse 11, 6330 Cham, Switzerland

Contents

Contributors

Vida Abedi Nutritional Immunology and Molecular Medicine Laboratory, Biocomplexity Institute of Virginia Tech, Blacksburg, VA, USA

Biomedical and Translational Informatics Institute, Geisinger Health System, Danville, PA, USA

Josep Bassaganya-Riera Nutritional Immunology and Molecular Medicine Laboratory, Biocomplexity Institute of Virginia Tech, Blacksburg, VA, USA

BioTherapeutics Inc., Blacksburg, VA, USA

David R. Bevan Department of Biochemistry, Virginia Tech, Blacksburg, VA, USA

Marion Ehrich Department of Biomedical Sciences and Pathobiology, Virginia Tech, Blacksburg, VA, USA

Raquel Hontecillas Nutritional Immunology and Molecular Medicine Laboratory, Biocomplexity Institute of Virginia Tech, Blacksburg, VA, USA

Biotherapeutics Inc., Blacksburg, VA, USA

Andrew Leber Biotherapeutics Inc., Blacksburg, VA, USA

Pinyi Lu Biotherapeutics Inc., Blacksburg, VA, USA

Nariman Noorbakhsh-Sabet Department of Neurology, University of Tennessee Health Science Center, Memphis, TN, USA

Nuria Tubau-Juni Nutritional Immunology and Molecular Medicine Laboratory, Biocomplexity Institute of Virginia Tech, Blacksburg, VA, USA

Meghna Verma Nutritional Immunology and Molecular Medicine Laboratory, Biocomplexity Institute of Virginia Tech, Blacksburg, VA, USA

Ramin Zand Department of Neurology, University of Tennessee Health Science Center, Memphis, TN, USA

Department of Neurology, Geisinger Medical Center, Danville, PA, USA

Nutritional Immunology and Molecular Medicine Laboratory, Biocomplexity Institute of Virginia Tech, Blacksburg, VA, USA

Victoria Zoccoli-Rodriguez Biotherapeutics Inc., Blacksburg, VA, USA

Chapter 1
Introduction to Accelerated Path to Cures and Precision Medicine in Inflammatory Bowel Disease

Josep Bassaganya-Riera and Raquel Hontecillas

Abstract Path to Cures provides a transformative perspective on the power of combining computational technologies, modeling, bioinformatics and machine learning approaches with nonclinical and clinical experimentation to accelerate the path to cures. In the following chapters, we will discuss the application of modeling technologies from target identification and validation, to nonclinical studies in animals to Phase I–III human clinical trials and post-approval monitoring. As a use case of successful integration of computational modeling and drug development, we will discuss the lanthionine synthetase C-like 2 (LANCL2) pathway and the development of LANCL2-based oral small molecule therapeutics for inflammatory bowel disease (IBD). From the application of docking studies to screen new chemical entities to the development of next-generation *in silico* human clinical trials.

Keywords Modeling · Drug development · Precision medicine · IBD · LANCL2

Overview

How many of us have shunned, rejected or ignored medication because, "the treatment is worse than the disease"? How many of us have taken drugs only to experience side effects as bad as our disease symptoms? Well, things are about to change. Breakthroughs in precision medicine and health and modeling-enabled drug development have the potential to change the experiences of those of us suffering from a wide range of human diseases. And, these changes will profoundly transform the global drug development landscape forever.

J. Bassaganya-Riera (✉) · R. Hontecillas
Nutritional Immunology and Molecular Medicine Laboratory, Biocomplexity Institute of Virginia Tech, Blacksburg, VA, USA

Biotherapeutics Inc., Blacksburg, VA, USA
e-mail: jbassaga@vt.edu

© Springer International Publishing AG, part of Springer Nature 2018
J. Bassaganya-Riera (ed.), *Accelerated Path to Cures*,
https://doi.org/10.1007/978-3-319-73238-1_1

1

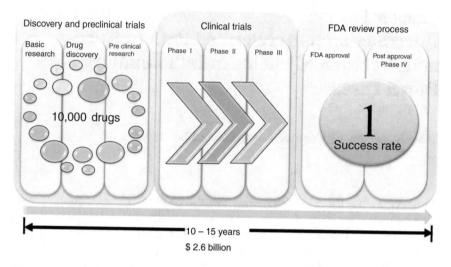

Fig. 1.1 Drug discovery and development processes

The drug approval process in the U.S. is very rigorous. The cost to develop a drug is about $2.6 billion and the timeline is 10–15 years to go from the beginning of animal testing to market. The chances of success are slim: one in 10,000. These facts and figures illustrate the inefficiency of the drug development process in the U.S. They highlight the stringency of the process for bringing a drug to market involving discovery, testing in animals and in humans through Phase I to III clinical trials, and post approval monitoring. The Food and Drug Administration (FDA) is the gatekeeper at each step of the process (Fig. 1.1). How can we: (1) improve the efficiency of the drug development process, and (2) develop safer and more effective drugs?

When engineers build airplanes, they do not use trial and error. Imagine what a waste that would be. They create computational models of airplanes before the real planes are built. The system is no longer trial and error, but an efficient, well-controlled and precise process that uses the power of modeling and simulation. Can biomedical researchers become the precision medicine engineers?

In adapting to the high rate of failure and the more than $2.6 billion total research and development (R&D) costs involved in the approval of a new drug or biologic (DiMasi et al. 2016), industry has turned increasingly to computational modeling at all levels, from modeling drug-receptor interactions to pharmacokinetic (PK) and pharmacodynamic (PD) modeling, to *in silico* clinical trials (Holford et al. 2010; Terstappen and Reggiani 2001). The U.S. FDA's Critical Path initiative has further stimulated the incorporation of mathematical modeling in drug discovery (FDA 2004; Vodovotz et al. 2017).

We have come a long way since Hippocrates proposed the medicinal value of nutrition and food. We celebrated Metchnikov and his discovery of macrophages. Monoclonal antibodies have paved the way for targeted pathway research.

The efficiency of sequencing has increased 10,000-fold and its costs have decreased exponentially. Yet, we lack a comprehensive, systems-wide, mechanistic understanding of massively and dynamically interacting systems such as the immune system.

Nutritional immunology approaches that are centered around the discovery of novel immunoregulatory pathways such as lanthionine synthetase C-like 2 (LANCL2) and nucleotide-binding oligomerization domain-like receptor X1 (NLRX1) and that combine computational modeling and experimentation have the potential to translate research findings into drug development programs that can yield safer and more effective therapeutics for human diseases. For example, the LANCL2 pathway was originally identified as a target of the naturally occurring compound abscisic acid (ABA) in the context of a nutritional immunology program, and it is now the core technology in the development of novel orally active small molecule therapeutics for inflammatory bowel disease (IBD) and other metabolic and autoimmune diseases.

"Let food be thy medicine and medicine be thy food." Hippocrates, the father of modern medicine portrayed as the paragon of the ancient Greek physician understood the value of food and medicine in an integrative and organic way. His statement is at the very core of the concept of nutritional immunology. Yet, only 8% of Americans adhere to/live by this ancient principal. For instance, only 8% of us eat or drink the 5–9 daily servings of fruits and vegetables necessary for food to be our medicine. Nonetheless, novel therapeutic targets identified in the course of nutritional immunology studies can become promising therapeutic targets for drug development. For instance, the identification of LANCL2 as the natural receptor for ABA (Sturla et al. 2009) has led to the investigation of additional ligands (Lu et al. 2012) and progression into oral therapeutic development for IBD targeting LANCL2 (Bissel et al. 2016).

LANCL2 Drug Development Use Case

LANCL2 has emerged as a therapeutic target for chronic inflammatory, metabolic and immune-mediated diseases such as IBD (Lu et al. 2014). LANCL2 is expressed in neutrophils, monocytes, and splenocytes, plus in colonic CD4+ T cells and epithelial cells in the colon. ABA, the natural ligand, binds to LANCL2, leading to elevation of cyclic adenosine 3,5′-monophosphate (cAMP) and activation of protein kinase A (PKA) followed by suppression of inflammation (Bassaganya-Riera et al. 2011). Simulating models can guide clinical plans, predict results of clinical trials, and suggest new therapeutics. Through this modeling and experimental validation, a novel LANCL2-based oral therapeutic for IBD was identified and is being developed (Bissel et al. 2016). We developed *N,N*-bis(benzimidazolylpicolinoyl)piperazine (BT-11), as a ligand for LANCL2. We developed libraries of billions of new chemical entities (NCEs) from molecular docking methods (Lu et al. 2012), which predicted affinities and binding sites to LANCL2. The modeling predictions were validated biochemically by surface plasmon resonance (SPR), *in vitro* assessments, in mouse models of IBD, and in safety studies in rats. These integrated computational and experimental validation studies led to identifying/optimizing BT-11 as our top LANCL2-binding compound for IBD (Carbo et al. 2016). BT-11 is being developed as a first-in-class oral therapeutic for treating IBD. A recent paper further substantiates the LANCL2 technology as a therapeutic target for IBD and reports our initial, nonclinical findings on BT-11, a small-molecule therapeutic that outperforms current drugs and INDs (Carbo et al. 2016). Our safety studies in rats and dogs demonstrate a benign safety profile for BT-11 at doses that are 100 times higher than its effective dose (Bissel et al. 2016) and a benign profile even at the limit dose of 1000 mg/kg. *In silico* clinical trial simulations in IBD patients treated with LANCL2 drugs and standard of care will be used to help guide the design of Phase II and III clinical studies for LANCL2-based drugs (Abedi et al. 2016). Therefore, combining the advanced computational methods and translational research is a necessary step toward accelerating the path to the clinic for new drug candidates.

Network Modeling for PD Biomarker Analyses Use Case

Addressing the complexity of the inflammatory and immune responses has carried out network modeling at the clinical level in the context of blunt trauma (Abboud et al. 2016; Almahmoud et al. 2015a, b; Brown et al. 2015; Namas et al. 2015, 2016b; Zaaqoq et al. 2014), spinal cord injury (Zaaqoq et al. 2014), IBD (Wendelsdorf et al. 2010) and Pediatric Acute Liver Failure (PALF) (Azhar et al. 2013). Computational modeling methodology has also been applied successfully in an effort to stratify trauma patients with regard to their degree of multiple organ dysfunction syndrome (MODS), based on multiple, early (within 24 h) assessments of circulating inflammatory mediators (Namas et al. 2016a). Principal component analysis (PCA), a data reduction technique, can identify the core variables in a dynamic, multi-dimensional dataset. From a cohort of 472 blunt trauma survivors described in several studies (Abboud et al. 2016; Almahmoud et al. 2015b; Brown et al. 2015; Namas et al. 2016b), two separate sub-cohorts of moderately to severely injured blunt trauma patients were studied. Multiple inflammatory mediators were assessed in serial blood samples. Similar approaches can be applied in IBD to define "patient-specific inflammation barcodes", in concert with analysis of dynamic networks. Inflammatory mediator data will be analyzed initially with standard statistical analyses. PCA combined with hierarchical clustering analysis enables segregation of patients based on inflammatory profiles, and this is correlated with clinical outcomes. These computational approaches can suggest interrelationships among inflammatory mediators and may suggest networks of PD biomarkers (Azhar et al. 2013).

In summary, to decrease its inefficiency, the drug development process ought to evolve more rapidly from the paradigm of experiment, data analysis, and interpretation, to a more integrative systems approach that incorporates formal computational

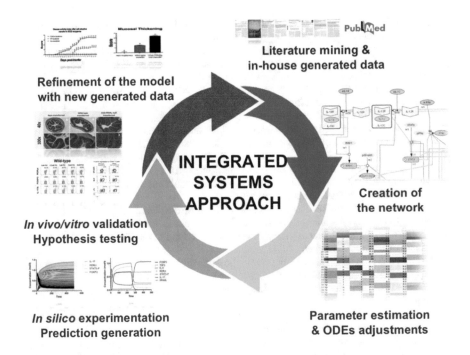

Fig. 1.2 Integrated systems approach

modeling and prediction (Fig. 1.2). The following chapters provide a window on the application of novel technologies from target identification and validation, to computer-aided drug discovery, to application of modeling in nonclinical testing, to network analyses for PD biomarker detection and validation, to the next generation of *in silico* human clinical trials. Additionally, throughout the chapters, we discuss mechanistic and data-driven modeling applied to precision medicine, strategies to connect complex, big and diverse data, models, and tools, strategies of multiscale modeling combining a variety of modeling technologies, such as equation-based modeling and agent based modeling, plus the application of advanced machine learning algorithms to nonclinical, translational and clinical R&D.

References

Abboud A et al (2016) Computational analysis supports an early, type 17 cell-associated divergence of blunt trauma survival and mortality. Crit Care Med 44:e1074–e1081. https://doi.org/10.1097/ccm.0000000000001951

Abedi V et al (2016) Phase III placebo-controlled, randomized clinical trial with synthetic Crohn's disease patients to evaluate treatment response. In: Arabnia H, Tran Q-N (eds) Emerging trends in computational biology, bioinformatics, and systems biology—systems & applications, 2ndedn. edn. Elsevier/MK

Almahmoud K et al (2015a) Impact of injury severity on dynamic inflammation networks following blunt trauma. Shock 44:105–109. https://doi.org/10.1097/shk.0000000000000395

Almahmoud K et al (2015b) Prehospital hypotension is associated with altered inflammation dynamics and worse outcomes following blunt trauma in humans. Crit Care Med 43:1395–1404. https://doi.org/10.1097/ccm.0000000000000964

Azhar N, Ziraldo C, Barclay D, Rudnick D, Squires R, Vodovotz Y (2013) Analysis of serum inflammatory mediators identifies unique dynamic networks associated with death and spontaneous survival in pediatric acute liver failure. PLoS One 8:e78202

Bassaganya-Riera J et al (2011) Abscisic acid regulates inflammation via ligand-binding domain-independent activation of PPAR gamma. J Biol Chem 286:2504–2516

Bissel P et al (2016) Exploratory studies with BT-11: a proposed orally active therapeutic for Crohn's disease. Int J Toxicol 35:521–529. https://doi.org/10.1177/1091581816646356

Brown D et al (2015) Trauma *in silico*: individual-specific mathematical models and virtual clinical populations. Sci Transl Med 7:285ra261

Carbo A, Gandour RD, Hontecillas R, Philipson N, Uren A, Bassaganya-Riera J (2016) An N,N-Bis(benzimidazolylpicolinoyl)piperazine (BT-11): a novel lanthionine synthetase C-like 2-based therapeutic for inflammatory bowel disease. J Med Chem 59:10113–10126. https://doi.org/10.1021/acs.jmedchem.6b00412

DiMasi JA, Grabowski HG, Hansen RW (2016) Innovation in the pharmaceutical industry: new estimates of R&D costs. J Health Econ 47:20–33. https://doi.org/10.1016/j.jhealeco.2016.01.012

FDA (2004) Innovation or stagnation: challenge and opportunity on the critical path to new medical products

Holford N, Ma SC, Ploeger BA (2010) Clinical trial simulation: a review. Clin Pharmacol Ther 88:166–182. https://doi.org/10.1038/clpt.2010.114

Lu P et al (2012) Computational modeling-based discovery of novel classes of anti-inflammatory drugs that target lanthionine synthetase C-like protein 2. PLoS One 7:e34643. https://doi.org/10.1371/journal.pone.0034643

Lu P, Hontecillas R, Philipson CW, Bassaganya-Riera J (2014) Lanthionine synthetase component C-like protein 2: a new drug target for inflammatory diseases and diabetes. Curr Drug Targets 15:565–572

Namas RA et al (2015) Insights into the role of chemokines, damage-associated molecular patterns, and lymphocyte-derived mediators from computational models of trauma-Induced inflammation. Antioxid Redox Signal 23:1370–1387. https://doi.org/10.1089/ars.2015.6398

Namas RA et al (2016a) Individual-specific principal component analysis of circulating inflammatory mediators predicts early organ dysfunction in trauma patients. J Crit Care 36:146–153. https://doi.org/10.1016/j.jcrc.2016.07.002

Namas RA et al (2016b) Temporal patterns of circulating inflammation biomarker networks differentiate susceptibility to nosocomial infection following blunt trauma in humans. Ann Surg 263:191–198. https://doi.org/10.1097/sla.0000000000001001

Sturla L et al (2009) LANCL2 is necessary for abscisic acid binding and signaling in human granulocytes and in rat insulinoma cells. J Biol Chem 284:28045–28057. https://doi.org/10.1074/jbc.M109.035329

Terstappen GC, Reggiani A (2001) In silico research in drug discovery. Trends Pharmacol Sci 22:23–26

Vodovotz Y et al (2017) Solving immunology? Trends Immunol 38:116–127

Wendelsdorf K, Bassaganya-Riera J, Hontecillas R, Eubank S (2010) Model of colonic inflammation: immune modulatory mechanisms in inflammatory bowel disease. J Theor Biol 264:1225–1239. https://doi.org/10.1016/j.jtbi.2010.03.027

Zaaqoq AM et al (2014) Inducible protein-10, a potential driver of neurally-controlled IL-10 and morbidity in human blunt trauma. Crit Care Med 42:1487–1497

Chapter 2
Computer-Aided Drug Discovery

Pinyi Lu[†]**, David R. Bevan**[†]**, Andrew Leber, Raquel Hontecillas,
Nuria Tubau-Juni, and Josep Bassaganya-Riera**

Abstract Computer-aided drug discovery has become an important part of the drug discovery process due to the reduced cost of computational methods and the increased availability of three-dimensional structural information. In this chapter, we compare structure-based and ligand-based modeling, and focus primarily on molecular docking and molecular dynamics simulations. In addition, we provide a broad overview of the application of computational methods in drug discovery and highlight some considerations in the application of molecular docking and molecular dynamics simulations. These approaches are particularly relevant in precision medicine because they have the potential to provide a detailed understanding of the molecular features that are essential for drug specificity and selectivity, thereby facilitating a comparison among population differences.

Keywords CADD · Structure-based modeling · Ligand-based modeling · Molecular docking · Molecular dynamics

[†]Pinyi Lu and David R. Bevan contributed equally to this work.

P. Lu (✉) · A. Leber
Biotherapeutics Inc., Blacksburg, VA, USA
e-mail: pinyi@biotherapeuticsinc.com

D. R. Bevan
Department of Biochemistry, Virginia Tech, Blacksburg, VA, USA

R. Hontecillas · J. Bassaganya-Riera
Nutritional Immunology and Molecular Medicine Laboratory, Biocomplexity Institute of Virginia Tech, Blacksburg, VA, USA

Biotherapeutics Inc., Blacksburg, VA, USA

N. Tubau-Juni
Nutritional Immunology and Molecular Medicine Laboratory, Biocomplexity Institute of Virginia Tech, Blacksburg, VA, USA

© Springer International Publishing AG, part of Springer Nature 2018
J. Bassaganya-Riera (ed.), *Accelerated Path to Cures*,
https://doi.org/10.1007/978-3-319-73238-1_2

Overview

Computational methods have been applied in drug discovery for many years, dating back to early applications of quantitative structure-activity relationships (QSAR) (Cherkasov et al. 2014). Computers are now applied as an important part of the drug discovery process for a number of reasons. One is that the skyrocketing costs of drug development have increased the appeal of low-cost computational methods. Another is the increased availability of three-dimensional structural information that can guide the characterization of drug targets as a way to streamline the discovery of potential drugs for those targets. As a result, computational methods typically are an integral component of drug discovery campaigns. These approaches are particularly relevant in precision medicine because they have the potential to provide a detailed understanding of the molecular features that are essential for drug specificity and selectivity, thereby facilitating a comparison among population differences. A number of reviews have appeared, which highlight different aspects of computer-aided drug discovery (Durrant and McCammon 2011; Ganesan et al. 2017; Leelananda and Lindert 2016; Mortier et al. 2015; Sliwoski et al. 2014). In this chapter, we will provide a broad overview of the application of computational methods in drug discovery and highlight some considerations in the application of molecular docking and molecular dynamics simulations.

Structure-Based Versus Ligand-Based Modeling

Computer-aided drug design is normally classified into ligand-based and structure-based modeling approaches (Fig. 2.1). When the structural information of targets is not available, ligand-based modeling approaches are often applied, including similarity searching, ligand-based pharmacophore mapping, and QSAR modeling. With advances in experimental techniques, such as high-throughput crystallography and nuclear magnetic resonance, an increasing number of three-dimensional structures of proteins have been determined. Structure-based modeling approaches are thus more widely used, mainly including molecular docking and structure-based pharmacophore mapping. Recently, the combined approaches of ligand and structure-based modeling

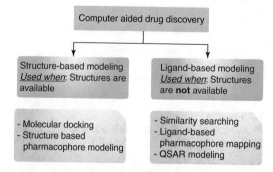

Fig. 2.1 Computer-aided drug discovery

have gained popularity. They integrate all biological and chemical information and reduce the limitations of each individual method (Drwal and Griffith 2013).

Ligand-Based Modeling Approaches

Similarity Searching

Similarity searching is based on the similar property principle, which searches compounds with similar characteristics to the active reference structure. These characteristics contain physicochemical properties (e.g. molecular weight and the logarithm of the partition coefficient, logP), as well as two and three-dimensional descriptors. For instance, two-dimensional properties are encoded as fingerprints, topological indices, and maximum common substructures. Similarity searching is simple, computationally inexpensive, and more effective than other methods for some targets, but it has bias towards input molecules and their structures (Drwal and Griffith 2013).

Ligand-Based Pharmacophore Modeling

The definition of pharmacophore from the International Union of Pure and Applied Chemistry (IUPAC) is "an ensemble of steric and electronic features that is necessary to ensure the optimal supramolecular interactions with a specific biological target and to trigger (or block) its biological response" (Wermuth et al. 1998). A pharmacophore model describes molecular features of ligands necessary to be recognized by receptors and explains how a common receptor site can be targeted by a set of structurally and functionally diverse ligands.

Typical pharmacophore features include hydrogen-bond donors and acceptors, positive and negative ionizable functional groups, inclusion/exclusion volume spheres, and ring centroids (Wang et al. 2008). These features could be located on the ligand itself or on the receptor. Therefore, pharmacophore modeling can be either ligand-based or structure-based, with the ligand-based pharmacophore approach being applied when no structural information of receptors is available.

The process for developing a ligand-based pharmacophore model includes the following steps: (1) Selection of a training set of ligands—Select a structurally diverse set of molecules, which should include both active and inactive compounds, in order to discriminate between molecules with and without bioactivity. (2) Molecular preparation—Generate low energy conformations for each of the selected molecules. (3) Molecular superimposition—Superimpose all combinations of the low-energy conformations of the molecules and abstract pharmacophore models. (4) Validation—Select preferred pharmacophore models for validation. Ligand-based pharmacophore modeling highly depends on the availability of a good training set of compounds manifesting the same binding mode (Lee et al. 2011).

Table 2.1 Classes of quantitative-structure activity relationships

Class	Description	Examples
0D	Descriptors derived from molecular formula	Molecular weight, Number and type of atoms
1D	Descriptors derived from substructure of molecules consisting of molecular fragments	Functional groups, Rings, Bonds, Substituents
2D	Descriptors derived from molecular graphs describing how atoms are bonded in a molecule	Total path count, Molecular connectivity indices
3D	Descriptors derived from geometrical representations of molecules	Molecular surface, Molecular volume, Electronic, Steric
4D	Descriptors derived from combination of atomic coordinates and sampling of conformations	Grid cell occupancy descriptors

Quantitative-Structure Activity Relationship Modeling (QSAR)

QSAR modeling uses knowledge of known active and inactive compounds to build a predictive regression or classification model. The QSAR predictors consist of physico-chemical properties or theoretical molecular descriptors of compounds and the QSAR response-variable is a biological function to quantify the potency of compounds. QSAR models summarize a mathematical relationship between chemical structures and biological activity in a dataset of compounds, which can then be used to predict the activities of new compounds (Cumming et al. 2013). Based on the dimensionality of molecular descriptors used, can be generally classified into five types (Table 2.1).

A QSAR model has the following form, and the error in the model includes model error and observational variability (Tropsha 2010).

Activity = f(physiochemical properties and/or structural properties) + error

Principal steps in QSAR studies include: (1) Selecting a curated chemical dataset. (2) Extracting structural/empirical descriptors. (3) Splitting data into training and test sets. (4) Constructing the QSAR model. (5) Validating the QSAR model. (6) Virtually designing ligands. (7) Selecting the best ligands. (8) Experimentally validating the compounds (Yousefinejad and Hemmateenejad 2015).

Structure-Based Modeling Approaches

Molecular Docking

Molecular docking is one of the most widely used structure-based modeling approaches, which can predict the binding mode of small molecules within the receptor binding sites and estimate the free energy of binding between ligands and receptors (Lengauer and Rarey 1996). In addition, the relative orientation of two interacting molecules may determine effects of the interactions. Interactions between biologically relevant molecules play important roles in signal transduction.

Thus, molecular docking is also able to predict the type of signals induced by molecular interactions (Ng et al. 2014).

The molecular docking process includes two steps. The first step is to predict conformations of the ligand and its position and orientation in the ligand binding site. The second step is to estimate the free energy of binding (Meng et al. 2011). Two key components, search algorithms and scoring functions, comprise these two steps. Search algorithms are rules and parameters used to predict the conformations of a binary complex. The first molecular docking program, DOCK, was published by Dr. Irwin Kuntz's group in 1982, which was created based on shape matching algorithms (Kuntz et al. 1982). Algorithms recently applied for molecular docking programs contain incremental construction, Monte Carlo simulations, simulated annealing, distance geometry, evolutionary programming (genetic algorithms), and tabu search (Dias and de Azevedo 2008).

In the field of molecular docking, scoring functions are approximate mathematical methods that are used to estimate binding affinities between targets and ligands. Scoring functions can be classified into three categories, force field-, empirical-, and knowledge-based (Macalino et al. 2015). Force field-based functions estimate binding affinity derived from non-bonded interactions in the target-ligand complex. Hydrogen bonds, solvation and entropy terms are also considered in the extensions of force field-based functions. Empirical-based functions estimate binding affinities by evaluating simple energy terms, including hydrogen bond, ionic interaction, hydrophobic effect and binding entropy. Knowledge-based functions derive binding energy from statistical analyses of a set of protein-ligand complex structures to obtain interatomic contact frequencies. Greater frequencies of occurrence represent more favorable interactions (Meng et al. 2011).

Structure-Based Pharmacophore Modeling

Compared with ligand-based pharmacophore modeling, structure-based pharmacophore modeling builds the pharmacophore models by identifying potentially important sites in protein active sites and translating them into pharmacophore features, thus it needs the structural information of protein targets (Drwal and Griffith 2013). Structure-based pharmacophore modeling can be divided into macromolecule-ligand-complex based and macromolecule-based methods, depending on whether or not structural information of ligands is used (Pirhadi et al. 2013). The former method derives the pharmacophore model from protein-ligand complexes, using the potential interactions between proteins and ligands, while the latter one builds pharmacophore models using protein active site information only.

Structure-based pharmacophore modeling does not need to solve problems faced by ligand-based pharmacophore modeling, such as ligand flexibility and molecular alignment. Furthermore, more detailed information related with entire interaction capability of the protein pocket can be utilized, including regions that ligands cannot reach, although this approach may increase computational costs and complicate the pharmacophore feature selection process due to a significantly larger number of potentially important sites. Structure-based pharmacophore modeling can be applied in virtual screening, hit to lead optimization, scaffold hopping, and multi-target drug design (Pirhadi et al. 2013).

Molecular Docking and Virtual Screening

Molecular Docking

The problem that molecular docking tries to solve is to identify the best orientation of a ligand that can bind to a specific protein. The principle of molecular docking is based on the lock-and-key theory (Fischer 1894), in which the ligand is considered as a key while protein is a lock. Thus, molecular docking can be also defined as finding the correct relative orientation of a "key" to open a "lock". However, because both ligands and proteins are flexible, the new analog, hand-in-glove, is more appropriate to be used for describing molecular docking. During the docking process, the ligand and the protein continue to adjust their conformations to achieve a stable complex with low free energy of binding (Wei et al. 2004).

There are two main classes of molecular docking methods, shape complementarity and simulation-based approaches. Shape complementarity-based approaches describe ligands and proteins using a set of features. Those features could contain molecular surface descriptors, in which solvent-accessible surface area is used to describe protein surface while the molecular surface of the ligand is described using a matching surface description (Halperin et al. 2002). Shape complementarity methods can screen a large number of ligands in a short time and predict whether they can bind to the protein binding site. However, this class of methods cannot model the dynamic changes in the conformations of ligands and proteins.

The three-dimensional structures of proteins of interest are required for molecular docking. Structures of proteins can be determined experimentally using techniques such as x-ray crystallography and NMR spectroscopy. The Protein Data Bank (PDB) is a worldwide repository of information for the three-dimensional structures of large biological molecules, which was established in 1971 and is currently managed by the Research Collaboratory for Structural Bioinformatics (RCSB) (Berman et al. 2000). The PDB, which is freely accessible, contains more than 125,000 biological macromolecular structures, including around 34,000 human protein structures and information of 22,000 ligands (www.rcsb.org).

When the experimentally determined protein structures are not available, homology modeling can be used to construct an atomic-resolution model of the protein based on its amino acid sequence. Three-dimensional structures of homologous proteins are used as templates and are required to perform homology modeling. The selection of the template structure and the alignment of the target sequence with the template sequence are the most important steps in the homology modeling procedure, because the quality of the homology model is largely dependent on the quality of template structure and its similarity to the target. Following template selection and sequence alignment, homology modeling involves model construction and model assessment (Marti-Renom et al. 2000).

Virtual Screening

Drug discovery and development is a complex, time-consuming, and expensive process. Developing a new drug takes an average of 10–15 years and costs almost \$3 billion, with an annual growth rate of 8.5% (DiMasi et al. 2016). Within the drug discovery process, the identification of effective hit compounds is a key step, from which these lead compounds can be optimized and developed into drug candidates. The challenge of identifying hits is mainly from the vast chemical space and large biological space (Dobson 2004). The estimated number of compounds that can be synthesized in theory is in the range of 10^{60} and the number of human proteins ranges from 250,000 to 10^6 (Lavecchia and Di Giovanni 2013). How to efficiently identify hit compounds from chemical and biological space is a big challenge for drug discovery. Advances in the fields of high-throughput screening (HTS) and combinatorial chemistry technologies in the early 1990s accelerated the hit identification process by testing large numbers of compounds simultaneously (Lavecchia and Di Giovanni 2013). However, compared with the size of chemical space, it is feasible to test only a small proportion of compounds due to the high cost caused by the low hit rate. In addition, such technologies are not suitable for all the drug targets. Virtual screening is a computational method to search libraries of small molecules *in silico* to identify potential hits and represents a complementary technology to HTS (Stahura and Bajorath 2004). In a comparative study to screen for inhibitors of tyrosine phosphatase-1B, virtual screening got a hit rate over 35% by yielding 354 hit compounds, 127 of which showed inhibitory activities. High throughput screening tested 400,000 compounds and only 81 of them showed effective inhibition (Sliwoski et al. 2014).

Virtual screening can be divided into two categories, structure-based virtual screening and ligand-based virtual screening, depending on availability of information related to the target and existing ligands (Lill 2013). When the structure of the target protein is obtainable, structure-based virtual screening can be applied, including molecular docking and structure-based pharmacophore mapping. When the structural information of targets is not available, different ligand-based virtual screening approaches are applicable depending on number and type of known ligands. When a single active ligand is known, similarity searching can be used. If a set of active ligands is available, ligand-base pharmacophore mapping can be applied. In case that information about inactive compounds is also obtainable, QSAR modeling is applicable. Details of each approach have been introduced in the previous section, "Structure-based versus Ligand-based Modeling".

Compound libraries are required to perform virtual screening, using either structure-based or ligand-based approaches. A compound library is a collection of chemicals that can be used in HTS or virtual screening. For each chemical, information such as name, structure, chemical and physical properties, bioactivities, or vendors, is stored in the compound library. Compound libraries can be obtained from public databases or commercial vendors (Lavecchia and Di Giovanni 2013). For example, ZINC is a free publicly-accessible database that was developed by University of California, San Francisco (UCSF), initially aiming to provide commercially-available compounds in ready-to-dock, 3D format for virtual screening (http://zinc15.docking.org/) (Irwin and Shoichet 2005). The latest version of ZINC contains over 120 million compounds with information, such as ligand annotation, availability for purchase, target, and biology annotation (Sterling and Irwin 2015). The current version is not only a compound database, but also a suite of tools for ligand discovery. PubChem is another widely-used public chemical database that stores information about chemical substances and their biological activities (Kim et al. 2016). PubChem was launched in 2004 and is a part of the Molecular Libraries Roadmap Initiatives of the US National Institutes of Health (NIH). As of June 2017, PubChem has stored over 234 million depositor-provided chemical substance descriptions, and data for more than 93 million compounds with over 1.2 million biological assay descriptions (https://pubchem.ncbi.nlm.nih.gov). In addition to public databases, there are also compound collections available from vendors, such as ChemBridge (http://www.chembridge.com) and eMolecules (https://www.emolecules.com/).

Application of Molecular Dynamics Simulations in Drug Discovery

General Considerations

The technique of molecular dynamics (MD) simulation has evolved into an effective method for querying the structure and dynamics of biomolecules at the atomistic level. MD can reveal details of molecular motion and interaction that are difficult, if not impossible, to study with current experimental techniques. The impact that computational methods such as MD have made to our understanding of complex chemical systems resulted in the awarding of the Nobel Prize in Chemistry to Karplus, Levitt, and Warshel in 2013. When MD simulations are applied in computational biology, it is typically either to analyze and extend experimental results or to design experiments. In either case, it is important to recognize that the most effective applications of MD arise from the combined application of experimental and computation. At the very least, MD simulations require a reliable three-dimensional starting structure, which is obtained preferably through X-ray diffraction because the models of protein structure that arise from crystallography typically are based upon the greatest number of constraints. Structures derived from NMR spectroscopy and comparative structure (homology) modeling, discussed above, also may be applicable for many MD studies.

MD simulation is a physics-based approach that relies on the development of a set of empirically-derived parameters to account for atomic interactions and motions. The combination of a functional form and a parameter set constitutes a molecular mechanics force field that can be used to calculate the energy of the molecular system. Several molecular mechanics force fields have been developed, including AMBER (Case et al. 2005; Lindorff-Larsen et al. 2010), CHARMM (Best et al. 2012), GROMOS (Oostenbrink et al. 2004), OPLS (Kaminski et al. 2001), and NAMD (Phillips et al. 2005). In addition, a variety of molecular modeling suites also have been developed, such as AMBER (Case et al. 2017), CHARMM (Brooks et al. 2009), NAMD (Phillips et al. 2005), and GROMACS (Hess et al. 2008; Pronk et al. 2013), in which some or all of these force fields can be used. Notably, these modeling suites and force fields are under continuous development, so it is important to identify the strengths and limitations of the most recent versions of the force fields so that the most appropriate one is used for a given application. An additional challenge when applying MD simulation in drug discovery is obtaining parameters for the ligand structures that are consistent with the force field used for the protein. Examples of modeling tools that assist in building and parameterizing ligand structures include PRODRG (Schüttelkopf and van Aalten 2004), CHARMM-GUI (Jo et al. 2008), and AmberTools (Case et al. 2017).

Applications of Molecular Dynamics in Generating Conformational Ensembles

As noted above, modeling methods in drug discovery can be grouped into structure-based drug design (SBDD) and ligand-based drug design (LBDD). For MD simulations, SBDD methods are used because a three-dimensional structure of the target is required. Several reviews of the application of MD simulation in drug discovery have appeared (Durrant and McCammon 2011; Ganesan et al. 2017; Leelananda and Lindert 2016; Mortier et al. 2015; Sliwoski et al. 2014), with one by DeVivo et al. (2016) being particularly relevant to the focus of our chapter. As discussed above, a challenge in molecular docking and virtual screening is the selection of the target structure against which to dock. If several crystal structures of a receptor (or enzyme) complexed with different ligands are available, these different structures with the ligand removed can be used for docking. A good example of a receptor for which many receptor-ligand structures are available is peroxisome proliferator activated receptor gamma (PPARγ) (Lewis et al. 2010). However, the availability of a large number of structures of a given receptor with several ligands is the exception, and even when several structures are available, they may not represent the conformational heterogeneity for a given receptor. MD simulation is an ideal tool to account for receptor flexibility and to generate ensembles of structures for docking and virtual screening. However, MD must be applied correctly to obtain an ensemble of structures that represents conformational space available to a given receptor. An early implementation of this approach was the relaxed complex scheme (RCS), in which structures are selected at a regular interval throughout the trajectory (Amaro et al. 2008; Lin et al. 2002, 2003). A notable outcome of this early study was that the binding modes similar to those observed in receptor-ligand crystal structures were those with the lowest free energies (Lin et al. 2003). Rather than docking to a relatively large number of receptor structures that are selected at regular intervals throughout an MD trajectory, another approach is to apply clustering methods to identify representative structures. An example of this approach was in screening for inhibitors of the lyase activity of DNA polymerase β (Barakat and Tuszynski 2011). In this case, the RCS scheme was applied as a way to account for dominant backbone dynamics, with 11 receptor structures being selected for docking based on RMSD conformational clustering. This approach enabled the identification of compounds with a higher predicted binding affinity for DNA polymerase β than pamoic acid, a known inhibitor. Another refinement of the RCS involved ensemble-based docking using MD simulations in which a biasing potential was applied to the ligand to restrict its conformational freedom (Campbell et al. 2014). Consequently, the sampling of the binding pocket conformations was restricted to those that were similar to conformations in the crystal structure and were thereby known to be relevant. These authors validated their method using cyclin-dependent kinase 2 and factor Xa. In another application of MD simulation to identify protein conformations for docking, Zhao and Caflisch (2015) developed a protocol in which MD was used to identify conformations distinct from those of crystal structures, especially in terms of the orientation of side chains in the binding pocket and/or the aperture of the pocket. By examining the interaction of known inhibitors or fragments of inhibitors with these structures, a single MD snapshot, rather than an ensemble of structures, was selected for high-throughput docking. Clearly, some of these sampling methods will bias the ensemble towards conformations that have been observed experimentally. In other cases, it may be desirable to sample a much broader range of conformations, and even all-atom MD simulations may not provide a complete sampling of the conformational landscape. In those cases, enhanced sampling methods such as metadynamics, replica-exchange, and temperature-acceleration may be applied (Abrams and Bussi 2014).

MD and Scoring

A well-recognized limitation in virtual screening is that the computed energy scores provided by these fast docking programs cannot reliably rank a set of ligands relative to their experimental measures of binding efficacy (e.g., binding affinity, IC_{50}). Often this limitation is not a serious problem when searching for novel hits from tens of thousands to hundreds of thousands of compounds in a database. While it is recognized that some potential binders may be overlooked, it is likely that some interesting compounds may be identified.

However, a number of researchers are working to improve the speed and accuracy of binding affinities, particularly as a way to improve the application of docking in the hit-to-lead optimization process. These methods are an application of MD simulation and include what are sometimes termed endpoint methods and alchemical methods (de Ruiter and Oostenbrink 2011). These methods attempt to include thermodynamic parameters and account for protein flexibility and the presence of water. The endpoint methods include molecular mechanics/Poisson-Boltzmann surface area (MM/PBSA) and molecular mechanics/generalized born surface area (MM/GBSA); the alchemical methods include free energy perturbation (FEP) and thermodynamic integration (TI). Although these methods have been around for several years, continued efforts are invested to improve their speed and accuracy (Abel et al. 2017; Wang et al. 2015). Because these methods for calculating free energy are relatively time-intensive, they are not routinely used in scoring poses in virtual screening of large databases. However, they may be particularly applicable in the hit-to-lead stage of drug discovery in which selected hits are examined more carefully. In particular, the MM/PBSA and MM/GBSA methods have been evaluated for their utility in post-docking analysis. In general, these methods improve the accuracy of calculating binding free energies, though improvements with these methods are target specific (Hou et al. 2011a, b; Sgobba et al. 2012; Virtanen et al. 2015).

Another approach that is applied for improving docked poses is performing MD of the protein-ligand complexes. As with the methods involving free energy calculations, this approach is time-intensive and is best applied to a limited number of complexes. MD of complexes should identify those that are energetically unfavorable, and it can also reveal subtle conformational changes in the protein and/or the ligand that result in a better binding pose. In one recent example, MD of small molecules bound to G-protein-coupled receptors (GPCRs) deepened the understanding of the most relevant binding poses as well as the GPCR-ligand interactions that appeared to be most important to binding (Tautermann et al. 2015). Another very recent development was the application of induced fit docking (i.e., MD of the ligand-receptor complex) in combination with metadynamics simulation, which was noted above as an enhanced sampling technique (Clark et al. 2016). This approach significantly improved the accuracy of protein-ligand poses across a large test data set.

Docking typically is done by defining the binding site of interest and restricting the docking search space to the defined pocket. Poses of ligands within the pocket are generated and scored without regard for the path through which ligands must pass to enter the binding site. Although MD simulations of the binding process are computationally intensive, researchers have begun to perform unbiased simulations to examine drug binding from the bulk solvent into the protein target. This current work is laying the foundation for future studies in which this approach may be practical for routine application in drug discovery. From these simulations, it may be possible to get thermodynamic and kinetic parameters, which would be useful in prioritizing ligands in a drug optimization pipeline. Examples of the application of unbiased MD to study drug binding include dasatinib and PP1 binding to Src kinase (Shan et al. 2011), binding of benzamidine to trypsin (Buch et al. 2011), binding of beta blockers and a beta agonist to the β_1- and β_2-adrenergic GPCRs (Dror et al. 2011), and binding of a transition state analogue to purine nucleoside phosphorylase (Decherchi et al. 2015).

MD to Discover Novel Binding Sites

Another application of MD simulation in drug discovery is in the identification of novel binding sites, such as allosteric or cryptic binding sites in proteins. Protein kinases are one class of proteins for which allosteric effectors may be beneficial. Targeting the ATP binding site is relatively non-specific and can lead to undesirable effects on enzymes other than those being targeted. A recent review summarizes the conformational plasticity of protein kinases and the search for more selective inhibitors based on their conformational dynamics revealed through MD simulations (Tong and Seeliger 2015). Similarly, a paper by Foda et al. (2015) describes the application of MD to studies of an allosteric network of amino acids within protein kinases that couple regulatory sites with ATP- and substrate-binding sites. This information provides additional, novel targets that can be used in drug discovery. Some other applications of MD simulations in identifying allosteric sites involve GPCRs (Ivetac and McCammon 2010), Ras proteins (Grant et al. 2011), and *Staphylococcus aureus* Sortase A (Kappel et al. 2012). A similar application of MD simulations to generate a conformational ensemble led to the discovery of a cryptic binding site in HIV integrase, into which known inhibitors were found to dock (Schames et al. 2004).

Success Stories in Computational Drug Discovery

As with most new technologies, unrealistic expectations were initially placed upon computational methods to revolutionize drug discovery. As the wave of euphoria passed, the trough of disillusionment followed (Schlick et al. 2011), and we are now at a point where we can evaluate contributions of computational methods to drug discovery more realistically. Discussions of success stories are challenging in that measures of success are not always definitive. Identifying examples in which the computational approaches described in this chapter initiated a drug discovery campaign that led to a drug currently being marketed is difficult due to the length of time required to bring a drug to market. Moreover, efficient drug discovery is a multi-pronged effort requiring scientists and physicians with a broad range of skills.

One survey of the application of computational methods in drug discovery was conducted by evaluating the literature using the term "virtual screening" (Ripphausen et al. 2010). From analysis of this survey, a list of 12 journals in which virtual screening papers appeared most often was compiled. Other data reported in this review were the frequencies with which given protein families were being targeted, and the potency distribution of hits towards the targets. Another review details successful applications of computer-aided drug discovery by providing a historical account of the development of 12 small molecules that are either in clinical trials or have been approved for therapeutic use (Talele et al. 2010).

With the recognition that computational methods can inform drug discovery, pharmaceutical companies have applied these approaches to varying extents in their drug development campaigns for many years. In one recent paper, the implementation and management of computer-based methods at Bayer HealthCare was described (Hillisch et al. 2015). The authors note that of approximately 20 of their new chemical entities that are being tested in Phase I trials, at least five originated from computational design. Some of the others also benefited to some extent from the application of computational methods. Another recent paper summarized the implementation and management of computer-aided drug design at Boehringer Ingelheim, though success stories were not discussed (Muegge et al. 2017).

Three specific examples of structure-based drug discovery in the pharmaceutical industry will be discussed in more detail. One of these involves the search for inhibitors of the Type I transforming growth factor β receptor kinase activity (TβRI). Researchers at Biogen applied virtual screening using a triarylimidazole compound as a starting point. Through screening of a commercially available database of about 200,000 compounds, they identified 87 compounds for their ability to inhibit autophosphorylation of TβRI (Singh et al. 2003). Their most effective compound was found to be identical to a compound that was identified by researchers at Eli Lilly using *in vitro* screening and examination of co-crystal structures (Sawyer et al. 2003; Sliwoski et al. 2014). The second example involves acetyl-CoA carboxylase (ACC), which is the rate limiting enzyme in fatty acid synthesis and serves as a potential target for treating obesity, diabetes, and fatty liver disease. Two isoenzymes denoted ACC1 and ACC2 exhibit different cellular localizations and functions. Inhibition of both isoforms by a lead compound denoted ND-630 was effective in reducing hepatic steatosis, improving insulin sensitivity, and modulating dyslipidemia in a rat model (Harriman et al. 2016). This compound was directed at a dimerization interface that was characterized by structure-based computer modeling specifically to identify compounds that could displace water molecules at the interface (Abel et al. 2017). Another example involves lanthionine synthetase C-like 2 (LANCL2), which has emerged as a new therapeutic target for treating inflammatory and immune-mediated diseases and diabetes (Fig. 2.2) (Lu et al. 2014). To study the function of LANCL2 through its structure, homology modeling of human LANCL2 was performed using the crystal structure of human LANCL1 as a templated (Lu et al. 2011). On the basis of the predicted structure of LANCL2, a structure-based virtual screening was used to screen ligands from public databases. NSC61610 was identified as the top ranked compound based on free energies of binding to LANCL2 (Lu et al. 2012). However, NSC61610, did not have the drug-like properties to develop it commercially. Surface plasmon resonance (SPR) studies were performed to confirm the direct binding of BT-11 and several other compounds to the LANCL2 protein. It has been shown that BT-11 outperforms current drugs and INDs for inflammatory bowel disease in a series of preclinical studies (Carbo et al. 2016).

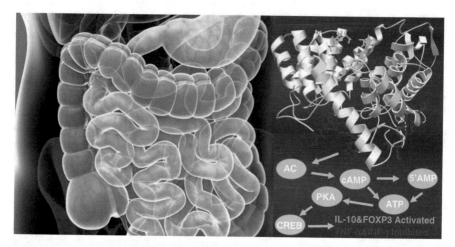

Fig. 2.2 LANCL2, a new therapeutic target for treating inflammatory bowel disease

Concluding Remarks

The volume of literature in recent years related to computer-aided drug discovery attests to the growing importance of these methods to accelerate drug development pipelines. In this chapter, we have focused primarily on molecular docking and molecular dynamics simulations, which are often applied in an integrated manner during the drug discovery process. Although virtual screening through molecular docking is a high-throughput process, the hits identified through virtual screening can be further refined using MD simulations. Moreover, MD simulations can provide a collection of structures that enables docking to be more effective in identifying compounds of interest. With future increases in computational power and the refinement of computational tools, the application of computers in drug discovery will continue to expand, which will ultimately lead to improvements in the efficiency of drug discovery and development.

References

Abel R, Wang L, Mobley DL, Friesner RA (2017) A critical review of validation, blind testing, and real- world use of alchemical protein-ligand binding free energy calculations. Curr Top Med Chem 17:2577–2585. https://doi.org/10.2174/1568026617666170414142131

Abrams C, Bussi G (2014) Enhanced sampling in molecular dynamics using metadynamics, replica-exchange, and temperature-acceleration. Entropy-Switz 16:163–199. https://doi.org/10.3390/e16010163

Amaro RE, Baron R, McCammon JA (2008) An improved relaxed complex scheme for receptor flexibility in computer-aided drug design. J Comput Aided Mol Des 22:693–705. https://doi.org/10.1007/s10822-007-9159-2

Barakat K, Tuszynski J (2011) Relaxed complex scheme suggests novel inhibitors for the lyase activity of DNA polymerase beta. J Mol Graph Model 29:702–716. https://doi.org/10.1016/j.jmgm.2010.12.003

Berman HM, Bhat TN, Bourne PE, Feng Z, Gilliland G, Weissig H, Westbrook J (2000) The Protein Data Bank and the challenge of structural genomics. Nat Struct Biol (7 Suppl):957–959. https://doi.org/10.1038/80734

Best RB, Zhu X, Shim J, Lopes PE, Mittal J, Feig M, Mackerell AD Jr (2012) Optimization of the additive CHARMM all-atom protein force field targeting improved sampling of the backbone phi, psi and side-chain chi(1) and chi(2) dihedral angles. J Chem Theory Comput 8:3257–3273. https://doi.org/10.1021/ct300400x

Brooks BR et al (2009) CHARMM: the biomolecular simulation program. J Comput Chem 30:1545–1614. https://doi.org/10.1002/jcc.21287

Buch I, Giorgino T, De Fabritiis G (2011) Complete reconstruction of an enzyme-inhibitor binding process by molecular dynamics simulations. Proc Natl Acad Sci U S A 108:10184–10189. https://doi.org/10.1073/pnas.1103547108

Campbell AJ, Lamb ML, Joseph-McCarthy D (2014) Ensemble-based docking using biased molecular dynamics. J Chem Inf Model 54:2127–2138. https://doi.org/10.1021/ci400729j

Carbo A, Gandour RD, Hontecillas R, Philipson N, Uren A, Bassaganya-Riera J (2016) An N,N-Bis(benzimidazolylpicolinoyl)piperazine (BT-11): a novel lanthionine synthetase C-like 2-based therapeutic for inflammatory bowel disease. J Med Chem 59:10113–10126. https://doi.org/10.1021/acs.jmedchem.6b00412

Case DA et al (2005) The Amber biomolecular simulation programs. J Comput Chem 26:1668–1688. https://doi.org/10.1002/jcc.20290

Case DA et al (2017) AMBER 2017. University of California, San Francisco

Cherkasov A et al (2014) QSAR modeling: where have you been? Where are you going to? J Med Chem 57:4977–5010. https://doi.org/10.1021/jm4004285

Clark AJ et al (2016) Prediction of protein-ligand binding poses via a combination of induced fit docking and metadynamics simulations. J Chem Theory Comput 12:2990–2998. https://doi.org/10.1021/acs.jctc.6b00201

Cumming JG, Davis AM, Muresan S, Haeberlein M, Chen H (2013) Chemical predictive modelling to improve compound quality. Nat Rev Drug Discov 12:948–962. https://doi.org/10.1038/nrd4128

de Ruiter A, Oostenbrink C (2011) Free energy calculations of protein-ligand interactions. Curr Opin Chem Biol 15:547–552. https://doi.org/10.1016/j.cbpa.2011.05.021

De Vivo M, Masetti M, Bottegoni G, Cavalli A (2016) Role of molecular dynamics and related methods in drug discovery. J Med Chem 59:4035–4061. https://doi.org/10.1021/acs.jmedchem.5b01684

Decherchi S, Berteotti A, Bottegoni G, Rocchia W, Cavalli A (2015) The ligand binding mechanism to purine nucleoside phosphorylase elucidated via molecular dynamics and machine learning. Nat Commun 6:6155. https://doi.org/10.1038/ncomms7155

Dias R, de Azevedo WF Jr (2008) Molecular docking algorithms. Curr Drug Targets 9:1040–1047

DiMasi JA, Grabowski HG, Hansen RW (2016) Innovation in the pharmaceutical industry: new estimates of R&D costs. J Health Econ 47:20–33. https://doi.org/10.1016/j.jhealeco.2016.01.012

Dobson CM (2004) Chemical space and biology. Nature 432:824–828. https://doi.org/10.1038/nature03192

Dror RO et al (2011) Pathway and mechanism of drug binding to G-protein-coupled receptors. Proc Natl Acad Sci U S A 108:13118–13123. https://doi.org/10.1073/pnas.1104614108

Drwal MN, Griffith R (2013) Combination of ligand- and structure-based methods in virtual screening. Drug Discov Today Technol 10:e395–e401. https://doi.org/10.1016/j.ddtec.2013.02.002

Durrant JD, McCammon JA (2011) Molecular dynamics simulations and drug discovery. BMC Biol 9:71. https://doi.org/10.1186/1741-7007-9-71

Fischer E (1894) Einfluss der configuration auf die wirkung der enzyme. Ber Dtsch Chem Ges 27:2985–2993

Foda ZH, Shan Y, Kim ET, Shaw DE, Seeliger MA (2015) A dynamically coupled alloste-ric network underlies binding cooperativity in Src kinase. Nat Commun 6:5939. https://doi.org/10.1038/ncomms6939

Ganesan A, Coote ML, Barakat K (2017) Molecular dynamics-driven drug discovery: leap-ing forward with confidence. Drug Discov Today 22:249–269. https://doi.org/10.1016/j.drudis.2016.11.001

Grant BJ, Lukman S, Hocker HJ, Sayyah J, Brown JH, McCammon JA, Gorfe AA (2011) Novel allosteric sites on Ras for lead generation. PLoS One 6:e25711. https://doi.org/10.1371/jour-nal.pone.0025711

Halperin I, Ma B, Wolfson H, Nussinov R (2002) Principles of docking: an overview of search algorithms and a guide to scoring functions. Proteins 47:409–443. https://doi.org/10.1002/prot.10115

Harriman G et al (2016) Acetyl-CoA carboxylase inhibition by ND-630 reduces hepatic steatosis, improves insulin sensitivity, and modulates dyslipidemia in rats. Proc Natl Acad Sci U S A 113:E1796–E1805. https://doi.org/10.1073/pnas.1520686113

Hess B, Kutzner C, van der Spoel D, Lindahl E (2008) GROMACS 4: algorithms for highly effi-cient, load-balanced, and scalable molecular simulation. J Chem Theory Comput 4:435–447

Hillisch A, Heinrich N, Wild H (2015) Computational chemistry in the pharmaceutical indus-try: from childhood to adolescence. ChemMedChem 10:1958–1962. https://doi.org/10.1002/cmdc.201500346

Hou T, Wang J, Li Y, Wang W (2011a) Assessing the performance of the MM/PBSA and MM/GBSA methods. 1. The accuracy of binding free energy calculations based on molecular dynamics simulations. J Chem Inf Model 51:69–82. https://doi.org/10.1021/ci100275a

Hou T, Wang J, Li Y, Wang W (2011b) Assessing the performance of the molecular mechanics/Poisson Boltzmann surface area and molecular mechanics/generalized Born surface area meth-ods. II. The accuracy of ranking poses generated from docking. J Comput Chem 32:866–877. https://doi.org/10.1002/jcc.21666

Irwin JJ, Shoichet BK (2005) ZINC—a free database of commercially available compounds for virtual screening. J Chem Inf Model 45:177–182. https://doi.org/10.1021/ci049714+

Ivetac A, McCammon JA (2010) Mapping the druggable allosteric space of G-protein coupled receptors: a fragment-based molecular dynamics approach. Chem Biol Drug Des 76:201–217. https://doi.org/10.1111/j.1747-0285.2010.01012.x

Jo S, Kim T, Iyer VG, Im W (2008) CHARMM-GUI: a web-based graphical user interface for CHARMM. J Comput Chem 29:1859–1865. https://doi.org/10.1002/jcc.20945

Kaminski GA, Friesner RA, Tirado-Rives J, Jorgensen WL (2001) Evaluation and reparametriza-tion of the OPLS-AA force field for proteins via comparison with accurate quantum chemical calculations on peptides. J Phys Chem B 105:6474–6487. https://doi.org/10.1021/jp003919d

Kappel K, Wereszczynski J, Clubb RT, McCammon JA (2012) The binding mechanism, mul-tiple binding modes, and allosteric regulation of Staphylococcus aureus Sortase A probed by molecular dynamics simulations. Protein Sci 21:1858–1871. https://doi.org/10.1002/pro.2168

Kim S et al (2016) PubChem substance and compound databases. Nucleic Acids Res 44:D1202–D1213. https://doi.org/10.1093/nar/gkv951

Kuntz ID, Blaney JM, Oatley SJ, Langridge R, Ferrin TE (1982) A geometric approach to macromolecule-ligand interactions. J Mol Biol 161:269–288

Lavecchia A, Di Giovanni C (2013) Virtual screening strategies in drug discovery: a critical review. Curr Med Chem 20:2839–2860

Lee CH, Huang HC, Juan HF (2011) Reviewing ligand-based rational drug design: the search for an ATP synthase inhibitor. Int J Mol Sci 12:5304–5318. https://doi.org/10.3390/ijms12085304

Leelananda SP, Lindert S (2016) Computational methods in drug discovery. Beilstein J Org Chem 12:2894–2718. https://doi.org/10.3762/bjoc.12.267

Lengauer T, Rarey M (1996) Computational methods for biomolecular docking. Curr Opin Struct Biol 6:402–406

Lewis SN, Bassaganya-Riera J, Bevan DR (2010) Virtual screening as a technique for PPAR mod-ulator discovery. PPAR Res 2010:861238. https://doi.org/10.1155/2010/861238

Lill M (2013) Virtual screening in drug design. Methods Mol Biol 993:1–12. https://doi.org/10.1007/978-1-62703-342-8_1

Lin JH, Perryman AL, Schames JR, McCammon JA (2002) Computational drug design accommodating receptor flexibility: the relaxed complex scheme. J Am Chem Soc 124:5632–5633

Lin JH, Perryman AL, Schames JR, McCammon JA (2003) The relaxed complex method: accommodating receptor flexibility for drug design with an improved scoring scheme. Biopolymers 68:47–62. https://doi.org/10.1002/bip.10218

Lindorff-Larsen K, Piana S, Palmo K, Maragakis P, Klepeis JL, Dror RO, Shaw DE (2010) Improved side-chain torsion potentials for the Amber ff99SB protein force field. Proteins 78:1950–1958. https://doi.org/10.1002/prot.22711

Lu P, Bevan DR, Lewis SN, Hontecillas R, Bassaganya-Riera J (2011) Molecular modeling of lanthionine synthetase component C-like protein 2: a potential target for the discovery of novel type 2 diabetes prophylactics and therapeutics. J Mol Model 17:543–553. https://doi.org/10.1007/s00894-010-0748-y

Lu P et al (2012) Computational modeling-based discovery of novel classes of anti-inflammatory drugs that target lanthionine synthetase C-like protein 2. PLoS One 7:e34643. https://doi.org/10.1371/journal.pone.0034643

Lu P, Hontecillas R, Philipson CW, Bassaganya-Riera J (2014) Lanthionine synthetase component C-like protein 2: a new drug target for inflammatory diseases and diabetes. Curr Drug Targets 15:565–572

Macalino SJ, Gosu V, Hong S, Choi S (2015) Role of computer-aided drug design in modern drug discovery. Arch Pharm Res 38:1686–1701. https://doi.org/10.1007/s12272-015-0640-5

Marti-Renom MA, Stuart AC, Fiser A, Sanchez R, Melo F, Sali A (2000) Comparative protein structure modeling of genes and genomes. Annu Rev Biophys Biomol Struct 29:291–325. doi:29/1/291 [pii]. https://doi.org/10.1146/annurev.biophys.29.1.291

Meng XY, Zhang HX, Mezei M, Cui M (2011) Molecular docking: a powerful approach for structure-based drug discovery. Curr Comput Aided Drug Des 7:146–157. doi:BSP/CCADD/E-Pub/000030 [pii].

Mortier J, Rakers C, Bermudez M, Murgueitio MS, Riniker S, Wolber G (2015) The impact of molecular dynamics on drug design: applications for the characterization of ligand-macromolecule complexes. Drug Discov Today 20:686–702. https://doi.org/10.1016/j.drudis.2015.01.003

Muegge I, Bergner A, Kriegl JM (2017) Computer-aided drug design at Boehringer Ingelheim. J Comput Aided Mol Des 31:275–285. https://doi.org/10.1007/s10822-016-9975-3

Ng HW et al (2014) Competitive molecular docking approach for predicting estrogen receptor subtype alpha agonists and antagonists. BMC Bioinformatics 15(Suppl 11):S4. https://doi.org/10.1186/1471-2105-15-S11-S4

Oostenbrink C, Villa A, Mark AE, van Gunsteren WF (2004) A biomolecular force field based on the free enthalpy of hydration and solvation: the GROMOS force-field parameter sets 53A5 and 53A6. J Comput Chem 25:1656–1676. https://doi.org/10.1002/jcc.20090

Phillips JC et al (2005) Scalable molecular dynamics with NAMD. J Comput Chem 26:1781–1802. https://doi.org/10.1002/jcc.20289

Pirhadi S, Shiri F, Ghasemi JB (2013) Methods and applications of structure based pharmacophores in drug discovery. Curr Top Med Chem 13:1036–1047

Pronk S et al (2013) GROMACS 4.5: a high-throughput and highly parallel open source molecular simulation toolkit. Bioinformatics 29:845–854. https://doi.org/10.1093/bioinformatics/btt055

Ripphausen P, Nisius B, Peltason L, Bajorath J (2010) Quo vadis, virtual screening? A comprehensive survey of prospective applications. J Med Chem 53:8461–8467. https://doi.org/10.1021/jm101020z

Sawyer JS et al (2003) Synthesis and activity of new aryl- and heteroaryl-substituted pyrazole inhibitors of the transforming growth factor-beta type I receptor kinase domain. J Med Chem 46:3953–3956. https://doi.org/10.1021/jm0205705

Schames JR, Henchman RH, Siegel JS, Sotriffer CA, Ni H, McCammon JA (2004) Discovery of a novel binding trench in HIV integrase. J Med Chem 47:1879–1881. https://doi.org/10.1021/jm0341913

Schlick T, Collepardo-Guevara R, Halvorsen LA, Jung S, Xiao X (2011) Biomolecular modeling and simulation: a field coming of age. Q Rev Biophys 44:191–228. https://doi.org/10.1017/S0033583510000284

Schüttelkopf AW, van Aalten DMF (2004) PRODRG: a tool for high-throughput crystallography of protein-ligand complexes. Acta Crystallogr D Biol Crystallogr 60:1355–1363

Sgobba M, Caporuscio F, Anighoro A, Portioli C, Rastelli G (2012) Application of a post-docking procedure based on MM-PBSA and MM-GBSA on single and multiple protein conformations. Eur J Med Chem 58:431–440. https://doi.org/10.1016/j.ejmech.2012.10.024

Shan Y, Kim ET, Eastwood MP, Dror RO, Seeliger MA, Shaw DE (2011) How does a drug molecule find its target binding site? J Am Chem Soc 133:9181–9183. https://doi.org/10.1021/ja202726y

Singh J et al (2003) Successful shape-based virtual screening: the discovery of a potent inhibitor of the type I TGFbeta receptor kinase (TbetaRI). Bioorg Med Chem Lett 13:4355–4359

Sliwoski G, Kothiwale S, Meiler J, Lowe EW Jr (2014) Computational methods in drug discovery. Pharmacol Rev 66:334–395. https://doi.org/10.1124/pr.112.007336

Stahura FL, Bajorath J (2004) Virtual screening methods that complement HTS. Comb Chem High Throughput Screen 7:259–269

Sterling T, Irwin JJ (2015) ZINC 15—ligand discovery for everyone. J Chem Inf Model 55:2324–2337. https://doi.org/10.1021/acs.jcim.5b00559

Talele TT, Khedkar SA, Rigby AC (2010) Successful applications of computer aided drug discovery: moving drugs from concept to the clinic. Curr Top Med Chem 10:127–141

Tautermann CS, Seeliger D, Kriegl JM (2015) What can we learn from molecular dynamics simulations for GPCR drug design? Comput Struct Biotechnol J 13:111–121. https://doi.org/10.1016/j.csbj.2014.12.002

Tong M, Seeliger MA (2015) Targeting conformational plasticity of protein kinases. ACS Chem Biol 10:190–200. https://doi.org/10.1021/cb500870a

Tropsha A (2010) Best practices for QSAR model development, validation, and exploitation. Mol Inform 29:476–488. https://doi.org/10.1002/minf.201000061

Virtanen SI, Niinivehmas SP, Pentikainen OT (2015) Case-specific performance of MM-PBSA, MM-GBSA, and SIE in virtual screening. J Mol Graph Model 62:303–318. https://doi.org/10.1016/j.jmgm.2015.10.012

Wang H, Duffy RA, Boykow GC, Chackalamannil S, Madison VS (2008) Identification of novel cannabinoid CB1 receptor antagonists by using virtual screening with a pharmacophore model. J Med Chem 51:2439–2446. https://doi.org/10.1021/jm701519h

Wang L et al (2015) Accurate and reliable prediction of relative ligand binding potency in prospective drug discovery by way of a modern free-energy calculation protocol and force field. J Am Chem Soc 137:2695–2703. https://doi.org/10.1021/ja512751q

Wei BQ, Weaver LH, Ferrari AM, Matthews BW, Shoichet BK (2004) Testing a flexible-receptor docking algorithm in a model binding site. J Mol Biol 337:1161–1182. https://doi.org/10.1016/j.jmb.2004.02.015

Wermuth CG, Ganellin CR, Lindberg P, Mitscher LA (1998) Glossary of terms used in medicinal chemistry (IUPAC Recommendations 1998). Pure Appl Chem 70:1129–1143

Yousefinejad S, Hemmateenejad B (2015) Chemometrics tools in QSAR/QSPR studies: a historical perspective. Chemom Intell Lab 149:177–204. https://doi.org/10.1016/j.chemolab.2015.06.016

Zhao H, Caflisch A (2015) Molecular dynamics in drug design. Eur J Med Chem 91:4–14. https://doi.org/10.1016/j.ejmech.2014.08.004

Chapter 3
Preclinical Studies: Efficacy and Safety

Nuria Tubau-Juni, Raquel Hontecillas, Marion Ehrich, Andrew Leber,
Victoria Zoccoli-Rodriguez, and Josep Bassaganya-Riera

Abstract Developing a successful drug from the discovery phase to its successful entry into the market costs an average of $3 billion dollars and 10–15 years. The continual goal of biopharma industry is to develop safer and more effective drugs. Therefore, performing *in vitro* and *in vivo* safety and efficacy studies is crucial in order to select the high potential drugs in preclinical phases and continue its development. Safety of a novel therapeutic agent must be rigorously analyzed and proven to prevent development of side effect in clinical testing. Thus, *in vivo* studies under GLP conditions and following FDA guidelines must be performed to advance towards IND approval. The utilization of several animal models is also required in order to test the efficacy of the compound. Efficacy studies offer the capacity to further dissect the mechanism of action of the novel compound and demonstrate its translational potential to humans.

Keywords Efficacy studies · Safety studies · Preclinical studies · Animal models · IBD

N. Tubau-Juni
Nutritional Immunology and Molecular Medicine Laboratory, Biocomplexity Institute of Virginia Tech, Blacksburg, VA, USA

R. Hontecillas (✉) · J. Bassaganya-Riera
Nutritional Immunology and Molecular Medicine Laboratory, Biocomplexity Institute of Virginia Tech, Blacksburg, VA, USA

Biotherapeutics Inc., Blacksburg, VA, USA
e-mail: rmagarzo@vt.edu

M. Ehrich
Department of Biomedical Sciences and Pathobiology, Virginia Tech, Blacksburg, VA, USA

A. Leber · V. Zoccoli-Rodriguez
Biotherapeutics Inc., Blacksburg, VA, USA

© Springer International Publishing AG, part of Springer Nature 2018
J. Bassaganya-Riera (ed.), *Accelerated Path to Cures*,
https://doi.org/10.1007/978-3-319-73238-1_3

Preclinical Studies: Safety

Safety studies are required by U.S. federal regulatory agencies to allay concerns about adverse health impacts of therapeutic agents before any new therapeutic agents are marketed. Determination of safety, which is the reciprocal of toxicity, results from evaluation of studies that examine toxicity or lack thereof. The most preliminary of experiments on safety examine toxicity to cultured cells; additional studies examine safety in live animals. Only the latter meet standards of the United States regulatory agency responsible for safety and efficacy of new therapeutic agents, the Food and Drug Administration (FDA). General references that provide comprehensive details on experiments needed to determine safety/toxicity for proposed new therapeutic agents are available (Faqi 2013; Hayes 2008; Ng 2004).

The importance of animal safety studies requires emphasis. Any individual or company who intends to manufacture a chemical for medicinal use is subject to federal regulatory guidelines. The FDA regulates drugs (prescription and non-prescription drugs), medical devices, food additives (including residues), and cosmetics. The FDA exists to protect the public and this requires toxicity testing in animals to assure that drugs used in the diagnosis and treatment of human diseases are safe. The initial testing requirements for safety (single dose studies in animals) are similar for any potential therapeutic agent. The studies described below are designed to meet toxicity testing guidelines of the Food and Drug Administration (FDA) (Lumpkin M and U.S. Food and Drug Administration 1995). To meet FDA requirements, these studies are to be done under Good Laboratory Practice (GLP) regulations, which prescribes that studies have established protocols, that experiments are done using standard operating procedures, and that data are available and subject to audit (FDA 2016). GLP studies are required if results are to be used as part of the submission for an Investigational New Drug Application (IND) from the FDA. FDA approval of the IND is required before evaluation of new drugs can be initiated in people.

Before determining safety in animals, it is common to do *in vitro* experiments to examine cytotoxicity of the test material to cultured cells. The Interagency Coordinating Committee on the Validation of Alternative Methods (ICCVAM) and the National Toxicology Program (NTP) Interagency Center for the Evaluation of Alternative Toxicological Methods (NICEATM) have provided guidelines so that the results of the *in vitro* experiment can be used to choose an appropriate dose for the first administration of a test compound to live animals (National Institute of Environmental Health Sciences et al. 2001). In addition such experiments follow recommendations of the National Research Council of the National Academies for Toxicity Testing in the twenty-first Century (National Research Council of the National Academies 2007), as *in vitro* testing can provide information on toxicity pathways and set the stage for targeted toxicity testing that increases potential that observation of animals used for *in vivo* experiments can provide information above and beyond general toxicity or safety.

In addition to the *in vitro* testing of the previous paragraph, pharmacokinetic/ toxicokinetic studies are also recommended (Faqi 2013; U.S. Department of Health and Human Services et al. 2010). Some of these experiments are also conducted *in vitro*, including capability to cross membranes, plasma protein binding, and capability to be biotransformed. Additional experiments involving dosing of animals with determination of test compound concentrations in samples (primarily blood) collected at various times after administration. These pharmacokinetic (PK) experiments provide information on how long the proposed drug may stay in the body, information particularly useful when designing repeated dose studies. PK are also useful for comparing effects on plasma concentrations when different routes of administration or different formulations of test agents are used. For example, comparison of drug concentrations after administration by the intravenous (IV) route, when absorption is not an issue, and by another route of administration that is expected to result in systemic effects but altered by factors relating to absorption (e.g., oral) can be used to determine the bioavailability of the test compound. These experiments compare maximal concentration and time the concentrations of drug can be therapeutically useful or safe. For these studies, plasma samples are taken at frequent intervals within hours after a single administration of the test agent (e.g., 0.2, 0.5, 1, 2, 4, 8, 12, 16, 24 h).

Single Dose Acute Toxicity Testing: These are safety studies conducted following FDA guidelines, including GLP (Center for Drug Evaluation and Research (CDER) 1996). Unless the proposed therapeutic agent is to be administered by another route, acute testing by the oral route is preferred by the FDA, as this usually provides systemic effects. These experiments are usually first done in both sexes of small rodents (rats, mice), but must also be done in a non-rodent mammalian species (e.g., rabbit, dog) to comply with the FDA guidelines. If sufficient test material is available, the rat is generally preferred over the mouse because behavioral changes are more easily quantitated (U.S. Department of Health and Human Services et al. 2007; U.S. Environmental Protection Agency 1998). Assuming that the preliminary experiment (described above) suggests a low order of toxicity, the maximum feasible dose would be the highest administered. (The FDA limit is 2 gm/kg, but it is possible that the maximal feasible dose will be lower based on solubility and proposed human exposure.) In order to assess dose-response relationships, two lower doses, including the dose that is expected to be applicable to people, will also be given. A minimum of five test rodents per dose per sex per group, including controls, meets the FDA requirements, but a group size of ten may be recommended if behavioral testing is included in the test protocol. As few as three test animals may be used for experiments in dogs if toxicity is expected to be unlikely based on the rodent studies. Animals are observed, including weekends, for 14 days after the single 24-h period in which the test compound was administered. They are weighed before dosing and weekly thereafter. Blood samples for clinical pathology are usually taken early after dosing (e.g., 8–24 h) and at time of sacrifice (14 days). If part of the experimental protocol, behavioral testing may be done before dosing and at intervals after dosing, such as 24 h, 7 days after dosing, and the day before sacrifice. In addition, animals are observed daily for general health and activity. All mortalities, clinical signs, time of onset, duration, and reversibility are recorded. Gross necropsies will be performed on all animals, including those sacrificed moribund, found dead, or terminated at 14 days.

Adult male and female rats (Long Evans, Wistar, or Sprague Dawley) at least 42 days old are generally preferred for use in the studies done in rodents. They are kept in quarantine prior to dosing. The three dosages are determined after review and consultation with the preparer of the product, and often have the highest dose based on availability and on the results of the preliminary experiments described above. Two lower dosages, equidistant logarithmatically apart, are also recommended to be given, with the lowest being that proposed as applicable to humans. A negative control group is also provided as untreated rats or those given only vehicle. The limit (highest) dose for testing is 2 gm/kg/day.

After the last behavioral determination, 3–5 rats per group are perfusion-fixed and organs removed, including lung, spleen, adrenal, liver, gastrointestinal tract, reproductive organs, kidney and brain. Histopathological evaluation uses paraffin-embedded sections stained with hematoxylin and eosin. Location and incidence of lesions, and descriptive information are used for pathological evaluation. A scoring system from 0–4 may be considered, especially if significant lesions are found.

Reports of studies done under GLP regulations contain information on test methods, experimental design, and results that provide information on individual animals as to effect of test compound on body weights, behavioral observations, and clinical and anatomical pathology. Summary data are provided, including statistical analysis. Also included are interpretative comments by board-certified professional personnel on the project (e.g., the toxicologist and the veterinary pathologist).

If sufficient test compound is not available to do the proposed GLP study following FDA Guidelines in rats, the studies may be conducted in mice. The procedures described above will be similar. They will also be relatively similar when non-rodent mammals are used as test species.

Repeated dose safety testing: As with acute testing, these experiments are done in two species, including a rodent (adult male and female rats or mice) and a non-rodent species (e.g., rabbit, dog). The usual route of administration is oral. These studies may last for 14 days (often termed subacute experiments), 90 days (sub-chronic experiments) or up to 2 years (chronic experiments). Of particular concern is stability of the test agent over the period of testing (U.S. Food and Drug Administration 2003). Three dosage groups plus a control group of n = 10 rodents are recommended to assure survival of a sufficient number of test animals for histopathological examination at the end of the study. This number may be reduced, especially if a 90-day test only includes a limit dose of 1 gm/kg/day. For the shorter experiments (14 days), test animals may be sacrificed at the end of dosing, with inclusion of a recovery group kept for an additional 7–14 days. Clinical and anatomical pathology endpoints are used for evaluation of safety/toxicity. For longer experiments, cohorts of experimental animals are sacrificed, with blood and tissue collected at intervals during the testing period (e.g., 30 and 60 days for a 90-day experiment). Information provided to the FDA in support of an IND is similar to that provided after acute testing. PK experiments may also be conducted at various times during repeated dose studies to determine if parameters such as rate of absorption, peak concentration, time to peak concentration, or rate of elimination change during the dosing period. Samples are taken as noted in the discussion above.

Preclinical Studies: Efficacy

In vitro cultures have been widely used for drug development. However, they are generally less predictive of improvement of clinical outcomes and tissue pathology than animal models of disease. Therefore, the efficacy of novel therapeutics must be also tested in animal models of disease prior to human testing. Animal models provide complex systems where drug absorption, chemical modifications and generation of metabolites, as well as systemic versus local actions of the drug candidate can be assessed. The most common animal model of disease for preclinical experimentation is the mouse. The mouse genome has 99% similarity with humans. Large-scale studies can be performed due to minimal size and cost. Thousands of knock-outs, knock-downs, protein tags, and other expression modifications can be performed due to optimized genetic engineering technology (Vandamme 2014). However, significant differences exist between mouse and human physiology. Pig models are a valid and under-utilized option due to interspecies similarities on immunity, anatomy, physiology and microbiome (Kobayashi et al. 2012). The choice of an animal model is not one-size-fits-all as multiple other factors such as natural susceptibility to disease, accuracy of symptoms, and disease pathogenesis could suggest that other species, such as hamsters, ferrets, dogs, or non-human primates, may be the better option. As a general rule, considerations of reduction, replacement, and refinement guide the use of animal models and result in using the lowest order species needed to address the scientific questions being considered.

Complex human diseases have a multifactorial etiology. For instance, Inflammatory Bowel Disease (IBD) is caused by multiple genetic modifications, presenting greater difficulty in being reproduced in animal models. In fact, IBD is caused by interaction of genetic defects, microbiota and the environment, resulting inexcessive and dysregulated immune responses. The choice of disease model is based on the aim of the study, the hypothesis, or the drug development stage. In some cases legacy drug development data using certain models is available and can be used to guide the development of new drugs. In other cases, robust conclusions on therapeutic efficacy require the use of several models to prove that the therapeutic efficacy is not model-dependent. In this chapter, IBD will be used as a use case to walk through the use of several animal models to assess the efficacy of a new drug.

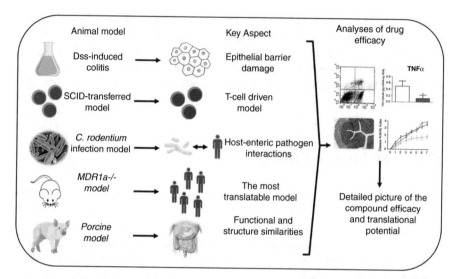

Fig. 3.1 Animal models most commonly used in efficacy studies for IBD in order to mimic the different aspects of the disease. DSS-induced colitis is a chemical-induced model that damages the epithelial barrier increasing gastrointestinal permeability. SCID-transferred model is a T-cell driven model based on T cell transfer into immunocompromised mice. C. rodentium infection model is also a T cell-driven model focused on the study of host-enteric pathogen interactions. MDRa1−/− model is characterized by the spontaneous development of a T-cell driven colitis and is the most translatable IBD model in mice. Porcine models report notable homology with human gastrointestinal tract and mucosal immune system becoming an adequate model for study of immune-mediated disease of human gut. While there is not a single model that shows better predictive value, demonstrating efficacy across models increases the confidence in human translation potential

Animal Models of IBD to Test Therapeutic Efficacy

Several animal models are available for the testing therapeutic efficacy in the context of autoimmune diseases, such as IBD. Each of the models has advantages and disadvantages that may make the individual model more suitable for specific stages of the drug development process or depending on the proposed mechanism of action or cell-specificity of the drug (Fig. 3.1).

DSS-Induced Colitis

Dextran sodium sulfate (DSS)-induced model of epithelial injury is an acute animal model for IBD that reproducibly promotes colitis and commonly used to study the mechanism of action and test efficacy of developing drugs (Viladomiu et al. 2013). DSS directly damages the epithelial barrier in the gastrointestinal tract, leading to increased intestinal permeability and induction of inflammation (Kitajima et al.

1999). Interestingly, DSS model provides controlled periods of active inflammation as well as certain intervals of recovery, mimicking the characteristic pathogenesis of IBD patients defined by acute periods of inflammation and followed by relapsing disease. The DSS-induced colitis model is generally utilized in the first steps of *in vivo* validation of the therapeutic efficacy of drugs.

Adoptive Transfer Model of Colitis

The adoptive transfer model is a T-cell driven model of colitis, in which CD4+CD45RB[hi] naive T cells are transferred to immunodeficient mice (e.g. SCID, RAG1−/−, RAG2−/−) (Steinbach et al. 2015; Viladomiu et al. 2013). SCID, RAG1−/− and RAG2−/− mice are characterized by the absence of functional lymphocytes due to an incomplete maturation of T and B cells. Therefore, approximately 5–8 weeks post-transfer, mice develop colitis and small bowel inflammation (Ostanin et al. 2009). In order to study specific T cell subsets, researchers may choose to combine the transfer of CD4+CD45RB[hi] T cells with CD4+CD45RB[lo]CD25+ regulatory T cells that will control the inflammation (Hontecillas and Bassaganya-Riera 2007; Leber et al. 2017c). This method can provide finer resolution to the mechanism of action of the molecular target by elucidating whether effector or regulatory arms of T cell responses are the mediators of the dominant effect of the drug.

Even though the specific mechanisms of IBD pathogenesis and etiology remain unknown, several studies have reported that a dysregulated immune response to the gut microbiome has a relevant role in the initiation of gastrointestinal inflammation in genetic susceptible patients (Ostanin et al. 2009). Thus, T-cell driven models, such as SCID- or RAG-transferred models, are the most appropriate to analyze the mechanisms by which this dysregulation occurs. These adoptive transfer models are frequently utilized to study drug modulation of T cell responses and allow the study of drug efficacy in initiation, maintenance and regulation of the disease.

Citrobacter rodentium *Infection Model of Colitis*

C. rodentium is an intestinal pathogen that colonizes the mucosa by means of attaching and effacing (A/E) lesions. Formation of A/E lesions is also a characteristic of human enteric pathogens enterophatogenic *Escherichia coli* (EPEC) and enterohaemorragic *E. coli* (EHEC). *C. rodentium* is a mouse-specific pathogen, therefore, it has emerged as one of the main mouse models for the study of EPEC and EHEC infections. Interestingly, development of colitis has been reported after C. *rodentium* infection; thus, the infection with this pathogen has been utilized as a model of IBD as well (Collins et al. 2014). *C. rodentium* is a T-cell driven model of IBD that is commonly being utilized to study the interaction of the host immune responses with enteric bacteria (Bhinder et al. 2013). This infectious model of colitis provides robust inflammatory responses stimulated by the generation of Th1 cells (Higgins et al. 1999) as well as the contribution of Th17 cell subsets (Symonds et al. 2009). This model has been used to study the IBD pathogenesis analyzing the mechanisms by which immune cells develop an excessive response to the microbiome reported in IBD. Similar to the T cell adoptive transfer model, *C. rodentium* model of colitis allows the study of the modulation of the immune response by a new therapeutic agent. More specifically, this model of colitis is developed as a result of a pathogen infection. Therefore, efficacy testing in *C. rodentium* infection is mainly focused on the induction of memory responses and tolerance of immune cells by the therapeutic agent.

MDR1a−/− Model of Colitis

MDR1a−/− mice are deficient in p-glycoprotein, an ATP-dependent small molecule efflux pump (Tanner et al. 2013) and develop spontaneous T-cell driven colitis when maintained under specific pathogen-free conditions (Tanner et al. 2013). MDR1a−/− mice exhibit similar pathogenesis to ulcerative colitis (UC) patients characterized by inflammation through the entire colon, mucosal thickening, immune cell infiltration, occasional ulcerations and dysregulated epithelial cell growth (Banner et al. 2004; Panwala et al. 1998). The MDR1a−/− model accurately recapitulates the role of host-microbiome interactions in the induction and perpetuation of gastrointestinal inflammation (Panwala et al. 1998), since colonic inflammation reported in MDR1a−/− diminishes after antibiotic treatment. The underlying mechanisms by which p-glycoprotein deficiency leads to inflammation of the large intestine with a similar pathology to human IBD remain unclear. The reported increase of epithelial permeability in MDR1a−/− mice (Collett et al. 2008) and the requirement of the microbiome to develop the pathogenesis suggests that exaggerated responsiveness to microbiome antigens translocated during impaired epithelial barrier function leads to mucosal inflammation and IBD symptoms.

In summary, studies performed in the MDR1a−/− mouse model of colitis are notable as they are the most translatable to human IBD pathogenesis. First, the etiology of MDR1a−/− histological findings in inflamed colon are highly similar to IBD patients. Second, the spontaneous development of the disease leads to a highly variable population of subjects with several degrees of disease severity, that would resemble the variability of a human population of IBD patients. Third, a correlation between the presence of some allelic variants of the MDR gene and increased predisposition to ulcerative colitis has been reported (Ho et al. 2006). Therefore, MDR1a−/− mice are a remarkably suitable IBD model for efficacy testing of new therapeutic compounds.

Pig as Model of Colitis

Pigs are less commonly utilized large animal models, although present remarkable advantages when studying inflammatory pathologies that target the gastrointestinal tract. Notable similarities have been reported in anatomy, physiology, structure, primary functions (e.g. absorption of nutrients) and mucosal immune system in the pig gastrointestinal tract compared to human gut (Jiminez et al. 2015). Moreover, when using pig models, a more comprehensive and systematic immune profiling can be performed since a substantial amount of immune cells can be isolated due to the size of pig organs (Viladomiu et al. 2013). Therefore, pig has become an ideal model to study human infectious and autoimmune diseases affecting the gastrointestinal mucosa. However, some disadvantages, including some differences reported to the human adaptive immune system, the requirement of specialized, highly regulated animal facilities in order to maintain the animals, increased costs, and low reproductive rate (Jiminez et al. 2015), are drawbacks to running large, primary studies in pigs.

Approacheds including bacterial-induced colitis (Hontecillas et al. 2005) or DSS-induced colitis (Bassaganya-Riera and Hontecillas 2006) have been utilized in order to reproduce IBD in the pig model. Interestingly, porcine model of colitis has been successfully utilized to test the efficacy of new therapeutic agents (Kim et al. 2010). Numerous studies performed in a pig DSS-induced colitis model reported the usage of this model to evaluate anti-inflammatory responses (Bassaganya-Riera and Hontecillas 2006; Ibuki et al. 2014; Kim et al. 2010) supporting the beneficial effects of using pigs as a model of IBD.

Cell-Specific Knockout Mice

The use of genetically modified animals as models of human diseases has significantly increased our understanding of the underlying molecular and signaling mechanisms of drug response. Cell-specific knockout mice enable detailed studies related to localization of action/activity of the drug and cell specificity of the mechanism of action. The selection of cell-specific knockout mice (CRISP or cre-lox based) depends highly on the cells involved in the pathogenesis of the disease.

In immune-mediated and infectious diseases of the gastrointestinal tract, immune and epithelial cells from the mucosa play a major role in disease development and pathogenesis. The use of epithelial (Wellman et al. 2017), myeloid cell (Hontecillas et al. 2011) as well as CD4+ T cell-specific knock out animals (Guri et al. 2010; Leber et al. 2017c) offer the capability to identify the cell specific role of novel, potential therapeutic targets for IBD. For instance, the specific deficiency of the Nucleotide oligomerization domain–like receptor X1 (NLRX1) in CD4+ T cells provided a powerful validation of the recent discovered role of NLRX1 in the regulation of T cells metabolism in a *C. rodentium* model of IBD (Leber et al. 2017c). In this study, the novel molecular mechanism by which NLRX1 regulates T cell activation, and therefore, ameliorates IBD severity in several mouse models of IBD was identified (Leber et al. 2017c).

In addition, the utilization of myeloid (Leber et al. 2016; Viladomiu et al. 2017) and CD4+ T (Leber et al. 2017a) cell-specific knock out in murine models of infection, provided new insights in the specific immune cell functions during host responses to the pathogen, as well as the identification and validation of novel potential therapeutic targets for infectious diseases.

Animal Models Used in Drug Development for IBD

Complex human diseases have a multifactorial etiology, and are not caused by only a single genetic modification, present more difficulty in being reproduced. In these cases, several models that mimic specific symptomatology of the diseases have been generated and are used to study the various aspects of the pathology.

The choice of animal model is based on the aim of the study, the hypothesis, or the drug development stage. In this chapter, IBD serves as a case study to walk through the use of several animal models in order to assess the efficacy of a new drug. Current IBD drugs in the market or in advanced clinical trials were previously tested in several animal models of IBD to assess their therapeutic efficacy before IND-enabling studies and clinical testing (Table 3.1).

Utilization of Computational Models in Preclinical Studies

The drug development process is extremely expensive (500 million to >2.6 billion per drug), complex and time-consuming, with a failure rate of 70–85% due to problems with efficacy and/or safety. Therefore, there is an urgent need to substantially improve the efficiency and reduce costs. To meet this need, biotech and biopharma companies ought to only invest on those compounds that are most likely to succeed in clinical trials and fail an unsuccessful investigation early in preclinical stages (Schmidt et al. 2013). Developing advanced computational systems that use preclinical data to predict outcomes of human clinical trials has the potential to help address the inefficiency problem in drug development. Indeed, the FDA's Critical Plan initiative has further stimulated the incorporation of mathematical modeling in drug discovery (FDA 2004).

Mechanistic modeling, such as the Systems of Ordinary Differential Equations (ODE) based model, is based on specific mechanistic hypotheses, and uses coupled mathematical equations and computer simulation to describe abstractions of complex biological systems. Mechanistic models that incorporate prior knowledge of data from *in vitro* and *in vivo* experiments, derive more insight from those experimental results, guide the design of following experiments to test new hypothesis, and translate the results into the context of human disease. However, the training and calibration of traditional mechanistic model is a tedious process that requires significant amount of time and expertise. In contrast, artificial Intelligence (AI)-based modeling is extremely efficient and adept at assigning weights and hidden, unknown interactions between parameters to accurately predict outcomes based on the data. Therefore, AI-based models are invaluable tools for usage in big datasets, such as *in* silico clinical trials, as it is further described in Chap. 5.

Mechanistic modeling and simulations are serving increasingly important roles in drug development and clinical trials (Vodovotz et al. 2017). The use of AI approaches at drug discovery and development stage is more limited. However, AI-based methods offer the potential to quickly adapt mechanistic models to the clinic into predictive and personalized datasets.

Table 3.1 Overview of IBD drugs currently in the market or in late stage development and the corresponding animal models used in preclinical efficacy studies

Therapeutic agent	Development stage	Description	Animal model	Translation to humans
GED0301	Phase 2–3 clinical testing in CD and UC. CD trials were stopped in October 2017	Smad7 antisense oligodeoxynucleotide. Facilitates Smad7 mRNA degradation (Ardizzone et al. 2016; Boirivant et al. 2006; Monteleone et al. 2015)	Chemical induced (TNBS and oxazolone) and adoptive transfer model (Boirivant et al. 2006)	Disease severity decrease in animal models is translated in initial human clinical trials (Monteleone et al. 2015) but Phase 2b/3 program for Crohn's was shut down in late 2017
TNFα blockers	Commercialized. Dominates 67% of the IBD therapeutic market	Biological therapies constituted by antibodies that block TNFα (Neurath 2017)	Adoptive Transfer model (Powrie et al. 1994) DSS-induced colitis (Kojouharoff et al. 1997; Lopetuso et al. 2013; Olson et al. 1995) and IL10−/− (Myers et al. 2003; Scheinin et al. 2003)	Variable degrees of efficacy. Some of the detected side effects in humans were not reported in animal models
α₄β₇ integrin antibodies	Phase 3 clinical testing (Vedolizumab)	Humanized monoclonal antibody that Inhibits T cell-endothelial cells interaction. Prevents T cells extravasation to the tissues (Singh et al. 2016; Soler et al. 2009)	SAMP1/Yit mice (MAdCAM-1) (Matsuzaki et al. 2005), Chronically colitic cotton-top tamarins (Hesterberg et al. 1996)	Treated patients reported higher remission rate than placebo (Singh et al. 2016). It mimics the reduced disease severity reported in animal models
S1P1 receptor modulators	Phase 3 clinical testing for UC and Phase 2 for Crohn's diseases (RPC1063)	Sphingosine-1-phosphate receptor modulators have been established as treatment for autoimmune diseases (Scott et al. 2016)	IL10−/− (KRP-203) (Song et al. 2008), DSS-induced colitis and adoptive transfer (FTY720) (Deguchi et al. 2006), TNBS-induced and adoptive transfer (RPC1063) (Scott et al. 2016)	Therapeutic effect of RPC1063 was validated in UC patients inducing clinical remission, clinical response and mucosal improvement (Sandborn et al. 2015)

Novel hybrid models that incorporate both mechanistic modeling and advanced machine learning methods from AI add a deeper layer of inference from large and multimodal datasets and provide powerful directions in drug development. A *Clostridium difficile (C. diff)* model of infection, combined traditional mechanistic modeling with AI algorithms. In this study, the generation of regulatory CD4+T cells that negatively correlates to infection recurrence and epithelium damage was identified from the computational model. Performing *in vivo* experimentation, Lanthionine Synthetase C-Like 2 (LANCL2) pathway was identified as a potential therapeutic mechanism to induce these beneficial effects on clinical outcomes, including mortality, recurrence and gut pathology. Finally, the activation of this pathway was validated to be an efficacious treatment in the AI-based model as well as the mouse model of infection (Leber et al. 2017b). Therefore, the utilization of computational approaches accelerated the *in vivo* preclinical experimentation refining the hypothesis and adding a second layer of validation.

AI-based models offer the capability to accelerate and streamline the drug development pipeline, in preclinical studies as well as clinical trials. For instance, a combined mechanistic and AI-based model accelerates the study of the therapeutic efficacy of the LANCL2 pathway in *in silico* clinical trials of IBD. The AI-based methodologies identify novel linkages between the biological parameters (i.e. immune cells, cytokines) and clinical outcomes, and therefore, offer the potential to adapt the mechanistic model of LANCL2 pathway to clinical and personalized dataset of *in silico* patients that can act as human avatars (See further details in Chap. 5).

Concluding Remarks

To advance in the development of novel therapeutic agents along the regulatory pipeline toward IND and clinical testing, safety and efficacy of the drug substance and drug product must be rigorously tested and proven. The essential goal in drug development is to generate safer and more effective drugs that can be commercialized to substantially improve the quality of life of the target patient population. *In vivo* studies in a rodent species and a non-rodent animal species are also performed under GLP conditions. Computational models have the potential to reduce, refine and replace the use of animals and improve the efficiency of nonclinical R&D. There is a need to develop predictive computational models of efficacy and safety to accelerate the drug development process.

References

Ardizzone S, Bevivino G, Monteleone G (2016) Mongersen, an oral Smad7 antisense oligonucleotide, in patients with active Crohn's disease. Therap Adv Gastroenterol 9:527–532. https://doi.org/10.1177/1756283X16636781

Banner KH, Cattaneo C, Le Net JL, Popovic A, Collins D, Gale JD (2004) Macroscopic, microscopic and biochemical characterisation of spontaneous colitis in a transgenic mouse, deficient in the multiple drug resistance 1a gene. Br J Pharmacol 143:590–598. https://doi.org/10.1038/sj.bjp.0705982

Bassaganya-Riera J, Hontecillas R (2006) CLA and n-3 PUFA differentially modulate clinical activity and colonic PPAR-responsive gene expression in a pig model of experimental IBD. Clin Nutr 25:454–465. https://doi.org/10.1016/j.clnu.2005.12.008

Bhinder G, Sham HP, Chan JM, Morampudi V, Jacobson K, Vallance BA (2013) The Citrobacter rodentium mouse model: studying pathogen and host contributions to infectious colitis. J Vis Exp:e50222. https://doi.org/10.3791/50222

Boirivant M et al (2006) Inhibition of Smad7 with a specific antisense oligonucleotide facilitates TGF-beta1-mediated suppression of colitis. Gastroenterology 131:1786–1798. https://doi.org/10.1053/j.gastro.2006.09.016

Center for Drug Evaluation and Research (CDER) (1996) Guidance for industry single dose acute toxicity testing for pharmaceuticals. https://www.fda.gov/downloads/Drugs/GuidanceComplianceRegulatoryInformation/Guidances/UCM079270.pdf. 2017

Collett A et al (2008) Early molecular and functional changes in colonic epithelium that precede increased gut permeability during colitis development in mdr1a(−/−) mice. Inflamm Bowel Dis 14:620–631. https://doi.org/10.1002/ibd.20375

Collins JW, Keeney KM, Crepin VF, Rathinam VA, Fitzgerald KA, Finlay BB, Frankel G (2014) Citrobacter rodentium: infection, inflammation and the microbiota. Nat Rev Microbiol 12:612–623. https://doi.org/10.1038/nrmicro3315

Deguchi Y, Andoh A, Yagi Y, Bamba S, Inatomi O, Tsujikawa T, Fujiyama Y (2006) The S1P receptor modulator FTY720 prevents the development of experimental colitis in mice. Oncol Rep 16:699–703

Faqi AS (ed) (2013) A comprehensive guide to toxicology in preclinical drug development. Elsevier, Amsterdam

FDA (2004) Innovation or stagnation: challenge and opportunity on the critical path to new medical products

FDA (2016) CFR—Code of Federal Regulations Title 21. Good laboratory practice for nonclinical laboratory studies. https://www.accessdata.fda.gov/scripts/cdrh/cfdocs/cfcfr/CFRSearch.cfm?CFRPart=58. 2016

Guri AJ, Mohapatra SK, Horne WT 2nd, Hontecillas R, Bassaganya-Riera J (2010) The role of T cell PPAR gamma in mice with experimental inflammatory bowel disease. BMC Gastroenterol 10:60. https://doi.org/10.1186/1471-230X-10-60

Hayes AW (ed) (2008) Principles and methods of toxicology, 5th edn. CRC Press, Boca Raton, FL

Hesterberg PE et al (1996) Rapid resolution of chronic colitis in the cotton-top tamarin with an antibody to a gut-homing integrin alpha 4 beta 7. Gastroenterology 111:1373–1380

Higgins LM, Frankel G, Douce G, Dougan G, MacDonald TT (1999) Citrobacter rodentium infection in mice elicits a mucosal Th1 cytokine response and lesions similar to those in murine inflammatory bowel disease. Infect Immun 67:3031–3039

Ho GT, Soranzo N, Nimmo ER, Tenesa A, Goldstein DB, Satsangi J (2006) ABCB1/MDR1 gene determines susceptibility and phenotype in ulcerative colitis: discrimination of critical variants using a gene-wide haplotype tagging approach. Hum Mol Genet 15:797–805. https://doi.org/10.1093/hmg/ddi494

Hontecillas R, Bassaganya-Riera J (2007) Peroxisome proliferator-activated receptor gamma is required for regulatory CD4+ T cell-mediated protection against colitis. J Immunol 178:2940–2949

Hontecillas R, Bassaganya-Riera J, Wilson J, Hutto DL, Wannemuehler MJ (2005) CD4+ T-cell responses and distribution at the colonic mucosa during Brachyspira hyodysenteriae-induced colitis in pigs. Immunology 115:127–135. https://doi.org/10.1111/j.1365-2567.2005.02124.x

Hontecillas R et al (2011) Immunoregulatory mechanisms of macrophage PPAR-gamma in mice with experimental inflammatory bowel disease. Mucosal Immunol 4:304–313. https://doi.org/10.1038/mi.2010.75

Ibuki M, Fukui K, Kanatani H, Mine Y (2014) Anti-inflammatory effects of mannanase-hydrolyzed copra meal in a porcine model of colitis. J Vet Med Sci 76:645–651

Jiminez JA, Uwiera TC, Douglas Inglis G, Uwiera RR (2015) Animal models to study acute and chronic intestinal inflammation in mammals. Gut Pathog 7:29. https://doi.org/10.1186/s13099-015-0076-y

Kim CJ, Kovacs-Nolan JA, Yang C, Archbold T, Fan MZ, Mine Y (2010) l-Tryptophan exhibits therapeutic function in a porcine model of dextran sodium sulfate (DSS)-induced colitis. J Nutr Biochem 21:468–475. https://doi.org/10.1016/j.jnutbio.2009.01.019

Kitajima S, Takuma S, Morimoto M (1999) Changes in colonic mucosal permeability in mouse colitis induced with dextran sulfate sodium. Exp Anim 48:137–143

Kobayashi E, Hishikawa S, Teratani T, Lefor AT (2012) The pig as a model for translational research: overview of porcine animal models at Jichi Medical University. Transplant Res 1:8. https://doi.org/10.1186/2047-1440-1-8

Kojouharoff G et al (1997) Neutralization of tumour necrosis factor (TNF) but not of IL-1 reduces inflammation in chronic dextran sulphate sodium-induced colitis in mice. Clin Exp Immunol 107:353–358

Leber A et al (2016) Modeling the role of lanthionine synthetase C-like 2 (LANCL2) in the modulation of immune responses to helicobacter pylori infection. PLoS One 11:e0167440. https://doi.org/10.1371/journal.pone.0167440

Leber A et al (2017a) Lanthionine synthetase C-like 2 modulates immune responses to influenza virus infection. Front Immunol 8:178. https://doi.org/10.3389/fimmu.2017.00178

Leber A, Hontecillas R, Abedi V, Tubau-Juni N, Zoccoli-Rodriguez V, Stewart C, Bassaganya-Riera J (2017b) Modeling new immunoregulatory therapeutics as antimicrobial alternatives for treating Clostridium difficile infection. Artif Intell Med 78:1–13. https://doi.org/10.1016/j.artmed.2017.05.003

Leber A et al (2017c) NLRX1 regulates effector and metabolic functions of CD4+ T cells. J Immunol 198:2260–2268. https://doi.org/10.4049/jimmunol.1601547

Lopetuso LR et al (2013) Locally injected infliximab ameliorates murine DSS colitis: differences in serum and intestinal levels of drug between healthy and colitic mice. Dig Liver Dis 45:1017–1021. https://doi.org/10.1016/j.dld.2013.06.007

Lumpkin M, U.S. Food and Drug Administration (1995) Guidance for industry. Content and formal of investigational new drug applications (INDs). Studies of drugs, including well-characterized, therapeutic, biotechnology-derived products. https://www.fda.gov/downloads/Drugs/GuidanceComplianceRegulatoryInformation/Guidances/UCM071597.pdf. 2017

Matsuzaki K et al (2005) In vivo demonstration of T lymphocyte migration and amelioration of ileitis in intestinal mucosa of SAMP1/Yit mice by the inhibition of MAdCAM-1. Clin Exp Immunol 140:22–31. https://doi.org/10.1111/j.1365-2249.2005.02742.x

Monteleone G et al (2015) Mongersen, an oral SMAD7 antisense oligonucleotide, and Crohn's disease. N Engl J Med 372:1104–1113. https://doi.org/10.1056/NEJMoa1407250

Myers KJ et al (2003) Antisense oligonucleotide blockade of tumor necrosis factor-alpha in two murine models of colitis. J Pharmacol Exp Ther 304:411–424. https://doi.org/10.1124/jpet.102.040329

National Institute of Environmental Health Sciences, National Institutes of Health, U.S. Public Health Service, Department of Health and Human Services (2001) Guidance document on using in vitro data to estimate in vivo starting doses for acute toxicity. https://ntp.niehs.nih.gov/iccvam/docs/acutetox_docs/guidance0801/iv_guide.pdf. 2017

National Research Council of the National Academies tNAP (2007) Toxicity testing in the 21st century: a vision and a strategy. Washington, DC. https://www.nap.edu/login.php?record_id=11970&page=https%3A%2F%2Fwww.nap.edu%2Fdownload%2F11970. 2017

Neurath MF (2017) Current and emerging therapeutic targets for IBD. Nat Rev Gastroenterol Hepatol 14:269–278. https://doi.org/10.1038/nrgastro.2016.208

Ng R (2004) Drugs from discovery to approval. John Wiley & Sons, Hoboken

Olson AD, DelBuono EA, Bitar KN, Remick DG (1995) Antiserum to tumor necrosis factor and failure to prevent murine colitis. J Pediatr Gastroenterol Nutr 21:410–418

Ostanin DV et al (2009) T cell transfer model of chronic colitis: concepts, considerations, and tricks of the trade. Am J Physiol Gastrointest Liver Physiol 296:G135–G146. https://doi.org/10.1152/ajpgi.90462.2008

Panwala CM, Jones JC, Viney JL (1998) A novel model of inflammatory bowel disease: mice deficient for the multiple drug resistance gene, mdr1a, spontaneously develop colitis. J Immunol 161:5733–5744

Powrie F, Leach MW, Mauze S, Menon S, Caddle LB, Coffman RL (1994) Inhibition of Th1 responses prevents inflammatory bowel disease in scid mice reconstituted with CD45RBhi CD4+ T cells. Immunity 1:553–562

Sandborn W et al (2015) 445 The TOUCHSTONE study: a randomized, double-blind, placebo-controlled induction trial of an oral S1P receptor modulator (RPC1063) in moderate to severe ulcerative colitis. Gastrointest Endosc 81:AB147. https://doi.org/10.1016/j.gie.2015.03.1236

Scheinin T, Butler DM, Salway F, Scallon B, Feldmann M (2003) Validation of the interleukin-10 knockout mouse model of colitis: antitumour necrosis factor-antibodies suppress the progression of colitis. Clin Exp Immunol 133:38–43

Schmidt BJ, Papin JA, Musante CJ (2013) Mechanistic systems modeling to guide drug discovery and development. Drug Discov Today 18:116–127. https://doi.org/10.1016/j.drudis.2012.09.003

Scott FL et al (2016) Ozanimod (RPC1063) is a potent sphingosine-1-phosphate receptor-1 (S1P1) and receptor-5 (S1P5) agonist with autoimmune disease-modifying activity. Br J Pharmacol 173:1778–1792. https://doi.org/10.1111/bph.13476

Singh H, Grewal N, Arora E, Kumar H, Kakkar AK (2016) Vedolizumab: a novel anti-integrin drug for treatment of inflammatory bowel disease. J Nat Sci Biol Med 7:4–9. https://doi.org/10.4103/0976-9668.175016

Soler D, Chapman T, Yang LL, Wyant T, Egan R, Fedyk ER (2009) The binding specificity and selective antagonism of vedolizumab, an anti-alpha4beta7 integrin therapeutic antibody in development for inflammatory bowel diseases. J Pharmacol Exp Ther 330:864–875. https://doi.org/10.1124/jpet.109.153973

Song J et al (2008) A novel sphingosine 1-phosphate receptor agonist, 2-amino-2-propanediol hydrochloride (KRP-203), regulates chronic colitis in interleukin-10 gene-deficient mice. J Pharmacol Exp Ther 324:276–283. https://doi.org/10.1124/jpet.106.119172

Steinbach EC, Gipson GR, Sheikh SZ (2015) Induction of murine intestinal inflammation by adoptive transfer of effector CD4+ CD45RB high T cells into immunodeficient mice. J Vis Exp. https://doi.org/10.3791/52533

Symonds EL, Riedel CU, O'Mahony D, Lapthorne S, O'Mahony L, Shanahan F (2009) Involvement of T helper type 17 and regulatory T cell activity in Citrobacter rodentium invasion and inflammatory damage. Clin Exp Immunol 157:148–154. https://doi.org/10.1111/j.1365-2249.2009.03934.x

Tanner SM, Staley EM, Lorenz RG (2013) Altered generation of induced regulatory T cells in the FVB.mdr1a−/− mouse model of colitis. Mucosal Immunol 6:309–323. https://doi.org/10.1038/mi.2012.73

U.S. Department of Health and Human Services, U. S. Food and Drug Administration, Center for Food Safety and Applied Nutrition (2007) Toxicological principles for the safety assessment of food ingredients. https://www.fda.gov/Food/GuidanceRegulation/GuidanceDocumentsRegulatoryInformation/IngredientsAdditivesGRASPackaging/ucm2006826.htm—TOC. 2017

U.S. Department of Health and Human Services USFaDA, Center for Drug Evaluation and Research (CDER), Center for Biologics Evaluation and Research (CBER), (2010) Guidance for industry M3(R2) nonclinical safety studies for the conduct of human clinical trials and marketing authorization for pharmaceuticals. https://www.fda.gov/downloads/Drugs/GuidanceComplianceRegulatoryInformation/Guidances/UCM073246.pdf. 2017

U.S. Environmental Protection Agency (1998) US EPA Guidelines for neurotoxicity risk assessment [FRL-6011-3]. https://www.epa.gov/sites/production/files/2014-11/documents/neuro_tox.pdf. 2017

U.S. Food and Drug Administration (2003) Toxicological principles for the safety assessment of food ingredients. Chapter IV.C.4.b Subchronic toxicity studies with non-rodents. https://www.fda.gov/food/guidanceregulation/guidancedocumentsregulatoryinformation/ingredientsadditivesgraspackaging/ucm078346.htm. 2017

Vandamme TF (2014) Use of rodents as models of human diseases. J Pharm Bioallied Sci 6:2–9. https://doi.org/10.4103/0975-7406.124301

Viladomiu M, Hontecillas R, Yuan L, Lu P, Bassaganya-Riera J (2013) Nutritional protective mechanisms against gut inflammation. J Nutr Biochem 24:929–939. https://doi.org/10.1016/j.jnutbio.2013.01.006

Viladomiu M et al (2017) Cooperation of gastric mononuclear phagocytes with Helicobacter pylori during colonization. J Immunol 198:3195–3204. https://doi.org/10.4049/jimmunol.1601902

Vodovotz Y et al (2017) Solving immunology? Trends Immunol 38:116–127. https://doi.org/10.1016/j.it.2016.11.006

Wellman AS et al (2017) Intestinal epithelial Sirtuin 1 regulates intestinal inflammation during aging in mice by altering the intestinal microbiota. Gastroenterology 153:772–786. https://doi.org/10.1053/j.gastro.2017.05.022

Chapter 4
From Nutritional Immunology to Drug Development

Meghna Verma, Raquel Hontecillas, Vida Abedi, Andrew Leber, Pinyi Lu, Nuria Tubau-Juni, and Josep Bassaganya-Riera

Abstract The chapter focuses on the key aspects of nutritional immunology that help accelerate the path to safer and more effective treatments. We review the complex network of interactions between diet, immune system, gut microbiome, and metabolism. We exemplify how advanced computational technologies such as artificial intelligence (AI) and machine learning methods combined with preclinical and translational models aid the discovery of novel immunoregulatory pathways. Further, we highlight nutritional immunology concepts translated to the development of novel therapeutics in immune mediated diseases such as IBD. Finally, we show how advancements in the field of nutritional immunology can be used to enable the development of predictive, preventive, and personalized path to safer and more effective drugs.

Keywords Nutritional immunology · Immunoregulatory pathways · Personalized nutrition · Precision medicine · Therapeutic targets · IBD

M. Verma · N. Tubau-Juni
Nutritional Immunology and Molecular Medicine Laboratory, Biocomplexity Institute of Virginia Tech, Blacksburg, VA, USA

R. Hontecillas · J. Bassaganya-Riera (✉)
Nutritional Immunology and Molecular Medicine Laboratory, Biocomplexity Institute of Virginia Tech, Blacksburg, VA, USA

Biotherapeutics Inc., Blacksburg, VA, USA
e-mail: jbassaga@vt.edu

V. Abedi
Nutritional Immunology and Molecular Medicine Laboratory, Biocomplexity Institute of Virginia Tech, Blacksburg, VA, USA

Biomedical and Translational Informatics Institute, Geisinger Health System, Danville, PA, USA

A. Leber · P. Lu
Biotherapeutics Inc., Blacksburg, VA, USA

© Springer International Publishing AG, part of Springer Nature 2018
J. Bassaganya-Riera (ed.), *Accelerated Path to Cures*,
https://doi.org/10.1007/978-3-319-73238-1_4

Overview

The dynamic interactions among nutrition, the gut microbiome and immune responses are initiated in early infancy when the development of early immune responses occurs. In prenatal and infancy stages, dietary nutrients shape oral tolerance to a wide array of commensal microbes and self-antigens.

Intake of glycoproteins in breast milk or infant supplements contribute to the development of tolerogenic dendritic cells and to oral tolerance (Zhou et al. 2010). Supplementation with long polyunsaturated fatty acids (PUFAs) including n-3 and n-6 PUFAs stimulate the differentiation of memory T cells and increase interleukin 10 (IL-10) production (Field et al. 2000) while decreasing the amount of IL-1β and tumor necrosis factor alpha (TNF-α) (Richard et al. 2016). Studies on the effect of maternal microbiome factors show that the microbiome colonization of the gut differs in breast-fed infants compared to formula-fed infants in that the former have fewer *Escherichia coli* and *Clostridium difficile* than the latter (Blümer et al. 2007; Penders et al. 2006). The beneficial and most abundant commensal clades of *Bifidobacterium* present in equal amounts in both groups, has a developed gene cluster for the metabolism of milk oligosaccharides which elucidates the involvement of gut microbiome in metabolic functions (Penders et al. 2006). Nutritional deficiencies of vitamins A, B-6, C, D, E and other micronutrients have been shown to affect both innate and adaptive cellular responses (Afacan et al. 2012). Malnutrition is detrimental to the functionality of the immune system, including predisposing immune cells toward an inflammatory state. An enteroaggregative *Escherichia coli* model of infection, demonstrated immunosuppressive effects of malnutrition at the cellular level (Philipson et al. 2013) while similar impaired immune responses were observed in a zinc deficient system (Bolick et al. 2014).

The interplay between diet, the gut microbiome, metabolism and the immune system is multidimensional *see* Fig. 4.1 and not limited to metabolites in symbiotic relationship, such as the production of short chain fatty acids (Smith et al. 2013). On the contrary, nutrient deprivation acts as a control mechanism to elicit significant changes in *Salmonella* genetic program to prevent invasion (Yurist-Doutsch et al. 2016). A balance between commensal and pathogenic species control the pathogenic expansion such as the inhibition of *Clostridium difficile* growth by commensal nontoxigenic *Clostridium* species (Buffie et al. 2015). The changes in microbiome not only contribute to the interaction between nutrition and immune system but recently, research into the gut-brain axis has unraveled links between nutrition, the microbiome and neurological disorders (Foster and McVey Neufeld 2013; Lyte 2013). The ingestion of dietary fat stimulated the cholecystokinin receptors and attenuated the inflammatory responses by the vagus nerve lowering levels of TNFα and IL6 cytokines (Luyer et al. 2005). Further, the appetite regulation hormone, leptin has been shown to influence Treg differentiation and migration of dendritic cells (Singh et al. 2014). Also, colonization with *Helicobacter pylori* increased leptin and ghrelin, and is associated with improved glucose homeostasis in mice with obesity and diabetes (Bassaganya-Riera et al. 2012a).

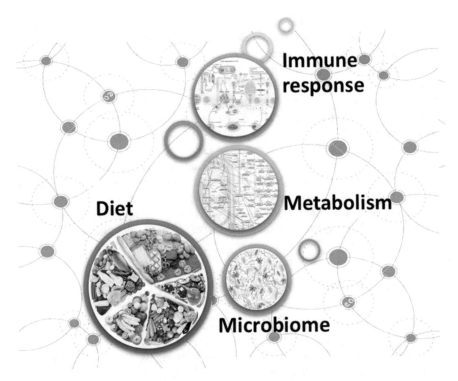

Fig. 4.1 Overview of the dynamic interactions that underlie nutritional immunology. The interplay between diet, gut microbiome, metabolism and immune response in the context of nutritional immunology is depicted

Nutritional Modulation of Immune-Mediated Diseases

The knowledge of the interaction between the immune system and diet naturally leads the focus to study the impact of nutritional therapeutics in immune mediated and infectious diseases in the gastrointestinal tract. Inflammatory bowel disease (IBD) is a debilitating disease characterized by chronic periods of intestinal inflammation (Neuman 2007; Zhang and Li 2014). Although the etiology of the disease remains unknown, it is believed that complex interaction between immune system, diet, environment, and genetic factors leads to disease development (Basson et al. 2016).The disrupted epithelial barrier function in susceptible patients allows for translocation of antigens from lumen to the intestinal layer thereby stimulating responses against commensal microbiota (Neurath 2014). An imbalance in the microbial composition compared to healthy individuals has been demonstrated towards IBD development (Ruemmele 2016). Studies examined modulation of the microbiome using probiotics and prebiotics to favor the growth of bacteria beneficial to the host. The bacterial community inside the gut digests resistant starches, generates lipids such as conjugated linoleic acid (CLA) and short chain fatty acids (SCFAs) such as acetate, propionate and butyrate which enhances intestinal health by promoting mucosal growth and intestinal permeability (Bassaganya-Riera et al. 2011b). Dietary supplementation with CLA in dextran sodium sulfate (DSS)-induced colitis mouse (Bassaganya-Riera et al. 2011a, 2012d; Evans et al. 2010) and pig models (Bassaganya-Riera and Hontecillas 2006) ameliorated disease severity and reduced leukocyte infiltration and colonic thickness (Bassaganya-Riera et al. 2011a). Additionally, VSL#3 and CLA probiotic treatment improved disease severity in DSS-induced colitis model (Bassaganya-Riera et al. 2012c). Thus, numerous studies demonstrated the relation between diet and IBD. Long term intake of n-3 poly saturated fatty acids (PUFAs) also exhibited reduced IBD development (Bassaganya-Riera and Hontecillas 2010; John et al. 2010). Overall, a comprehensive understanding of the nutritional quality of dietary interventions modulating the components of gut microbiome and their downstream effects on mucosal immune responses is crucial for use of precision medicine path for treating diseases of the lower gastrointestinal tract.

Important Immunometabolic Pathways

The metabolic activity of cells of the immune system plays a key role in activation, differentiation and proliferation processes. There is a clear demarcation between metabolic pathways involved in effector, regulatory and memory T cells. The resting naïve T cells, upon activation and when differentiating towards an effector T cell lineage, shift from a catabolic to an anabolic state (Newton et al. 2016). The effector T cells utilize aerobic glycolysis to meet the high-energy demand of cytokine production and control of their post-transcriptional phenotype (Chang et al. 2013; Michalek et al. 2011; Xu et al. 2016). This process is mainly driven by the glycolytic-lipogenic pathway associated with glutamine oxidation which fuels the mitochondrial oxidative phosphorylation through the tricarboxylic acid (TCA) cycle [2016]. The effector cells show enhanced fatty acid synthesis which is needed for their growth (O'Neill et al. 2016). In contrast to effector T cells, regulatory T (Treg) cells, both thymus derived and peripherally induced regulatory T cells have similar metabolic properties as that of memory T cells, in which they rely on glucose derived lipogenesis and fatty acid oxidation respectively [2016]. Cells that participate in the innate immune response have characteristic metabolic properties. For instance inflammatory and activated myeloid cells utilize aerobic glycolysis (Everts et al. 2014), while downregulating the oxidative pathways (Johnson et al. 2016). The activation of the innate immune response is inhibited by fatty acid transport or alteration in the TCA cycle. Innate immune cell activation is inhibited by fatty acid transport pathways or alteration in the TCA cycle control (Cipolletta et al. 2012). There are key metabolite sourced signaling pathways such as the protein kinase alpha (Akt) that has been identified as a strong repressor of differentiation into the Treg cell phenotype (Haxhinasto et al. 2008). The Akt pathway demonstrated the potential to negatively regulate the induction of the Treg cells (Haxhinasto et al. 2008).

The activation of lanthionine synthetase C-like 2 (LANCL2) pathway has been shown to exert anti-inflammatory effects (Lu et al. 2012). Further, other pathways including the nucleotide oligomerization domain like receptor X1 (NLRX1) (Allen et al. 2011) or peroxisome proliferator-activated receptor gamma (PPAR γ) pathways (Carbo et al. 2013b; Cipolletta et al. 2012) have the potential to play a role in the maintenance or disruption of anti-inflammatory environment. Additionally, the CX3C chemokine receptor 1 (CX3CR1) axis has been shown to contribute towards glucose and insulin control pathways (Lee et al. 2013), however contradictory results have also been reported (Shah et al. 2015). Further, it is important to note that CX3CR1 is a phenotypic marker for the regulatory macrophages involved in maintaining the intestinal homeostasis along with response to diseases such as IBD (Johnson et al. 2016; Lee et al. 2013). Nutritional therapeutics can be used to modulate the above immunometabolic pathways. Oleic acid, a common dietary component found in high concentration within *Parabacteroides* has been shown to reduce the disease severity of intestinal inflammation in murine models of colitis and chronic colitis (Kverka et al. 2011). The metabolic processes specific to certain immune cell subsets can be targeted pharmacologically to manipulate the immune system. It is important to note that multi-omic data integration through advanced computational modeling can facilitate a comprehensive understanding of how dietary and microbial factors modulate the metabolic axis to influence the gut immune responses (Verma et al. 2016).

Nutrition-Based Therapeutics Used to Treat Gastrointestinal Diseases

The identification of targets in the immune system that can be modulated via nutritional or metabolic factors lies at the core of nutritional immunology. Specifically, natural compounds that target anti-inflammatory pathways has been a major focus in recent years.

The peroxisome proliferator-activated receptor gamma (PPAR γ) is a transcription factor with anti-inflammatory properties shown to ameliorate IBD severity (Hontecillas et al. 2011). Activation of PPAR γ presents an anti-inflammatory mechanism (Bassaganya-Riera and Hontecillas 2006; Bassaganya-Riera et al. 2004; Hontecillas et al. 2009, 2011), which is involved in improving the intestinal health, and has numerous natural exogenous ligands that have been identified. It is also involved in lipid metabolism, and is a regulator glucose homeostasis (Picard and Auwerx 2002). It suppresses the pro-inflammatory pathways including the STAT, AP-1 and NF-κβ pathway and promotes CD4+ T cell plasticity towards regulatory subsets (Carbo et al. 2013b). Using virtual screening numerous nutritional ligands for PPAR γ have been identified including conjugated linoleic acid (CLA). Supplementation with CLA has been shown to suppress colitis in pig models wherein CLA suppressed colonic inflammation and upregulated colonic PPAR γ expression (Bassaganya-Riera and Hontecillas 2006). Oral administration of CLA in patients with mild to moderate Crohn's disease (a clinical manifestation of IBD), resulted in a significant reduction in the Crohn's disease activity index as well as reduced proinflammatory cytokine production including interferon gamma (IFN γ), tumor necrosis factor alpha (TNF α), and interleukin 17 (IL17) (Bassaganya-Riera et al. 2012b). Furthermore, probiotics including VSL#3, a mixture composed of four strains of lactobacilli, three strains of bifidobacteria and *Streptococcus thermophiles*, produces CLA locally, which resulted in PPAR γ dependent anti-inflammatory effects during experimental colitis in mice (Bassaganya-Riera et al. 2012c). Punicic acid (PUA), also known as trichosanic acid, is another naturally occurring conjugated triene fatty acid that has been demonstrated as a potential PPAR γ agonist (Bassaganya-Riera et al. 2011c). PUA is found in high concentration in the seeds of pomegranate and it has been demonstrated that oral PUA administration ameliorated glucose concentration (Hontecillas et al. 2009). Similar to CLA, PUA suppressed the expression of inflammatory cytokines against IBD (Bassaganya-Riera et al. 2011a). Resistant starch (RS), soluble corn fiber and inunlin have also been shown to ameliorate the clinical disease and prevent inflammatory lesions in animal model of IBD via activation of PPAR γ (Bassaganya-Riera et al. 2011b). The protective effects of RS involve production of short chain fatty acids (SCFA) (Bassaganya-Riera et al. 2011b), which is important due to the decreased amounts of healthy microbiota and SCFA reported in IBD patients, although, the immunomodulatory effects of RS remain unknown.

Another naturally occurring compound, abscisic acid (ABA), a phytohormone which is also naturally present in fruits and vegetables has been demonstrated to have anti-diabetic and anti-inflammatory properties (Guri et al. 2010a, b; Zocchi et al. 2017). The United States dietary survey suggests that about 92% of the human population might have deficient intake of ABA (Zocchi et al. 2017). The loss of PPAR γ in T cells eliminated anti-inflammatory efficacy of ABA against experimental IBD which demonstrated that ABA mediated activation of anti-inflammatory pathways (Guri et al. 2010a). Interestingly, ABA targets the pro-inflammatory pathways such as NF-$\kappa\beta$ (Guri et al. 2010b) in absence PPAR γ from T cells. Lanthionine synthetase C-like 2 (LANCL2) receptor has been identified as a mammalian receptor for ABA (Zocchi et al. 2017). ABA is shown to bind LANCL2 and activate LANCL2/cAMP initiated signaling and LANCL2 catalytic functions (Bassaganya-Riera et al. 2011d; Sturla et al. 2009). *In silico* studies performed on a model structure of LANCL2 suggested a potential ABA binding domain on LANCL2 which was later verified experimentally (Lu et al. 2011). The binding of ABA to LANCL2 acted independently of PPAR γ (Bassaganya-Riera et al. 2011d). As part of its downstream effects, LANCL2 has been shown to modulate the cyclic adenosine monophospahate (cAMP) responsive element binding protein (CREB)-dependent pathways (Sturla et al. 2009), the translocation of insulin independent stimulation of glucose transporter 4 (GLUT4) in adipocytes to the surface membrane (Bruzzone et al. 2012). LANCL2 also regulates the serine threonine protein kinase Akt pathway by mTOR (Zeng et al. 2014). Computational *in silico* approaches such as molecular docking identified potential ligands of LANCL2 (Lu et al. 2012). The results demonstrated that the lead compound NSC61610 engaged LANCL2 in an cAMP dependent manner and ameliorated experimental IBD by favoring Treg responses (Lu et al. 2012). Thus, the finding driven by a nutritional study related to the anti-inflammatory effect ABA lead to the discovery of LANCL2 pathway and, integrated *in silico* computational methods accelerated the path toward discovering novel nutritional ligands.

NLRX1 is a negative regulator in the nucleotide-binding domain and leucine rich repeat (NLR) family of cytosolic pattern recognition receptors. NLRs have been demonstrated to be involved in several infectious and immune mediated diseases such as IBD (Lu et al. 2015; Leber et al. 2017b). The loss of NLRX1 has been shown to increase the disease severity due to increase in populations of inflammatory cells in mouse model of DSS induced colitis (Leber et al. 2017b). Further, its role has been investigated in a *H. pylori* infection, such that NLXR1 deficiency lead to the clearance of the bacterium (Philipson et al. 2015). *In silico* approaches such as molecular docking was used to investigate the structure of NLRX1 and punicic acid (PUA) was identified as a natural ligand for NLRX1 (Lu et al. 2015). Further, the regulatory role of PUA was demonstrated on a mouse model of colitis which was shown to be NLRX1 dependent (Lu et al. 2015). Thus, PUA binds PPAR γ (Bassaganya-Riera et al. 2011a) as well as a NLRX1 (Lu et al. 2015), thereby demonstrating its anti inflammatory properties. Further, PUA has also been shown to have antidiabetic effects (Hontecillas et al. 2009; Vroegrijk et al. 2011).

Computational Approaches to Discover New Therapeutics: A Precision Medicine Initiative

With the advent of sequencing technologies, gene/immune profiling, and -omics analysis, patient related health outcomes and electronic health records have generated massive amounts of data. These data include different cellular effects, disease mechanisms and phenotypic differences across various populations. Many of these large-scale clinical and omics data sets are available for further analysis, allowing researchers to specifically look for diagnostic or disease biomarkers that help physicians to make better diagnoses and design therapeutics which are specific for a given patient.

Accelerating the Path to Precision Medicine

The total cost of developing a new drug based on the traditional pipeline remains high, ranging from $3 billion to up to $30 billion according to records for the years 2006–2014 (Schuhmacher et al. 2016; Wooden et al. 2017). The average time span between discovery and marketing is 15 years. Thus, high cost, a long time to use in patients, and low throughput are the combined challenges that stem from the traditional approaches to drug development. This drives us towards a need to use big data-driven approaches.

To accelerate the development of safer and more effective therapeutics for auto-immune diseases one should understand the immune system in health and in disease. The immune system is not defined linearly by simple signaling pathways but is a complex giant jigsaw puzzle or a Meccano game which includes an integrated response stemming from the complex interactions between genes, protein-protein interactions, and interactions with the environment (Gardy et al. 2009; Verma et al. 2016; Vodovotz et al. 2016). Computational modeling and advanced machine learning techniques are a crucial need for analyzing the complex network of interactions between nutrients, the gut microbiome, the metabolome, and the immune system (Allison et al. 2015). The big-data driven computational pipeline includes the identification of biomarkers by analysis of newly collected samples or *in silico* analysis of data derived from big data repositories (Wooden et al. 2017). Further stratification of patient datasets in the design of new clinical trials can be performed by screening them for relevant biomarkers and assigning the appropriate personalized therapy. This is termed 'basket trials' (Redig and Jänne 2015). These 'trials' are a new form of clinical trial design wherein the molecular marker predicts response to a therapeutic independent of the disease pathology (Redig and Jänne 2015). Furthermore, mathematical models and data driven approaches such as machine learning derived design of *in silico* clinical trials can help make predictions regarding the effect of drugs in a specific patient population and help better design the actual clinical trials needed before drugs are marketed (see Fig. 4.2).

The Nutritional Immunology and Molecular Medicine Laboratory (NIMML) (www.nimml.org) combined with modeling immunity to enteric pathogens (https://miep.org) has successfully implemented modeling approaches in the study of mucosal immune responses against infectious diseases. The work involved using computational modeling integrated with data analytics, informatics enabled by high performance computing to build information processing representation of the mucosal immune system. Some of the key projects involved developing an agent based model for diseases such as inflammatory bowel disease (Wendelsdorf et al. 2010), ordinary differential equation based models for *Clostridium difficile* (Leber et al. 2015), and multiple projects studying the gastric immune response to *Helicobacter pylori* (Carbo et al. 2013a, 2014; Leber et al. 2016), see Fig. 4.3.

Effective computational technologies including mathematical models such as equation based, agent based approaches, and machine learning models aid new hypothesis driven research and provide predictive solutions (Vodovotz et al. 2016). The equation based models rely on ordinary differential equation based models (Leber et al. 2016; Verma et al. 2017). This is in contrast to agent based models (Carbo et al. 2013a; Wendelsdorf et al. 2010), which use agents to represent the units in biological processes and each agent follows a set of rules and has unique properties. These properties include location, movement and interaction with other agents. The study of complex nutritional immunology models requires an understanding of the immune system dynamics at different spatio-temporal scales ranging from molecular to tissue level, population level and time ranging from nano-seconds to years. The *ENteric Immunity SImulator* multiscale model (*ENISI-MSM*) is one of the first agent-based simulators for the enteric immune system (Mei et al. 2014). It is based on ordinary differential

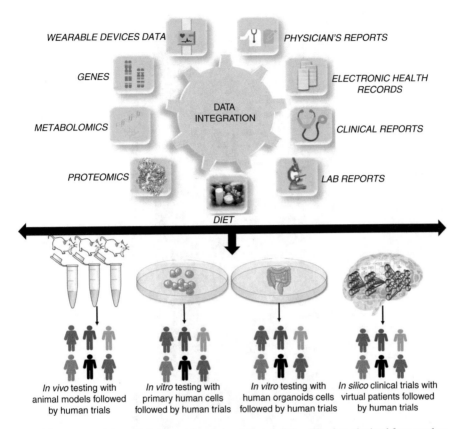

Fig. 4.2 Integrated data driven approaches to precision medicine. The data obtained from modern approaches such as proteomics and metabolomics along with gene signatures obtained from high sequencing methods when integrated with the electronic health records, clinical data, and patient reported outcomes can help in designing more precise treatments. The output data can be used in parallel to perform (1) in vivo testing in animal models which can be used in progress towards phase I clinical research, (2) in vitro testing in human organoid systems, and (3) in silico clinical trials with advanced machine learning models of the virtual clinical patients leading towards an accelerated path to personalized treatment of disease

equation (ODE) and partial differential equations, integrated as one unit within the agent based model (ABM). The *ENISI-MSM* can simulate signaling pathways, cytokine diffusions, cell movements and tissue lesions. Although the tool was designed from modeling mucosal immune responses which can simulate up to 10^9 of them using high performance computing (Mei et al. 2014; Wendelsdorf et al. 2012), it can be adapted to develop the representation of the host-microbiota-nutrient interactions (Verma et al. 2016). Currently, such approaches are still underdeveloped in the field of nutritional immunology and, as proposed in *Goals in Nutrition Science 2015–2020* (Allison et al. 2015), a mechanistic and big data driven approach towards understanding host nutrient microbiota interactions has an enormous potential to predict the outcomes of the nutrient-microbiota-metabolome-immune system axis.

Fig. 4.3 The models of immunity developed at the Nutritional Immunology and Molecular medicine laboratory since 2010–2017

Computational Models Used for Precision Medicine Discovery Measures Endpoints of Immune Mediated and Inflammatory Diseases

The data driven approaches have been integrated to build advanced machine learning algorithms for design of personalized diets based on capturing human diversity and variation (Zeevi et al. 2015). We developed an *in silico* clinical trial pipeline utilizing artificial intelligence based methods, to evaluate the efficacy of innovative *Clostridium difficile* (*C. diff*) treatments (Leber et al. 2017a). The study developed a computational pipeline which integrated preclinical model effects of treatments, evaluated host immunological measures and generated clinical scores for *C. diff* infections (Leber et al. 2017a). The model pipeline included a previously calibrated ordinary differential equation of *C. diff* infection (Leber et al. 2015), which was recalibrated to fit preclinical model data of the *C. diff* treatments, collected from their corresponding clinical trials that were in advanced clinical stages. The chosen clinical trial studies consisted of parameters of population level data with adverse effects, recurrent episodes of infection and length of hospital stay (Leber et al. 2017a). The recalibrated model outputs for each treatment was translated into clinical prediction scores using advanced machine learning algorithms. The pipeline predicted the effectiveness of LANCL2 activation, as a novel immune based treatment against *C. diff* infections (Leber et al. 2017a). Another *in silico* clinical trial model utilized the advanced machine learning method for use in the approach to treatment of Crohns Disease (Abedi et al. 2016). A group of 10,000 virtual patients was created by multi-model data integration. Longitudinal data from clinical trial as well as biological insights from field experts were used to create the virtual population. Experimental data from pre-clinical studies along with published results from clinical trial studies were used in the design of *in silico* clinical trial. This study provided novel non-intuitive way of hypothesis generation with regard to treatment using novel investigational therapeutics (Abedi et al. 2016). The virtual patients provided an avenue to explore the treatment options in the simulation set up. Thus, creating human avatars based on clinical data can further enhance the *in silico* clinical trials and pave the way to more successful studies, and thus reduced costs and time, in drug development. To achieve these goal, electronic health records and claims data from large institutions can provide valuable resource and further refine these predictive models.

In other studies, the authors developed a computational model which worked as a personalized nutrition predictor (Zeevi et al. 2015). The predictor was an advanced machine learning model which integrated the data from 800 people. Data input included their gut microbiome, results of a blood test, and a food diary among other parameters. Results predicted an accurate postprandial glucose response. The model was validated using an independent cohort and helped design personalized diets to lower the glycemic response (McDonald et al. 2016). The study involved tailoring diets of 26 people on the basis of their microbiome sample, blood sample and anthropometrics. The study was well acclaimed (Zeevi et al. 2015), and provided evidence that altering the diet, based on the person's personalized health, can produce specific and predictable health benefits (Zeevi et al. 2015). This is an example that demonstrates the enormous potential of big data to improve human health by aiding the design of an approach to personalized medicine.

The developed *in silico* pipelines (Abedi et al. 2016; Leber et al. 2017a) put forth a novel platform which has the potential to accelerate the development of novel immunoregulatory treatments for infectious and immune mediated diseases. At NIMML, a rich source of ordinary differential equations, agent based models and pipelines with integrated machine learning algorithms and artificial intelligence technologies, (see Fig. 4.3), combined with advanced data analytics capabilities can lead to development of a personalized, predictive and personalized tools. The integration of health data including the physiological parameters of a patient, electronic health records, clinical reports and lab reports, physician records, data from food diaries, wearable technologies data can be used to create *in silico* patients. The *in silico* patients will be used as avatars to computationally, (1) formulate descriptive analysis regarding the current status of the patient, (2) determine diagnostic analysis to study the responses in the past and, (3) test the personalized predictive medicines and, (4) evaluate the predictive scenarios to analyze the future condition of the patient and make recommendation for the future care.

Concluding Remarks

Nutritional immunology approaches integrated with advanced computational modeling and preclinical/translational experimentation centered around investigating novel immunoregulatory mechanisms, have the potential to translate new insights on immunoregulatory mechanisms controlled by nutrition or metabolism into a safer drug development pipeline. These novel drug discovery and development approaches have the potential to create predictive, preventive and personalized pipelines that identify new targets based on nutritional immunology insights and yield safer and more effective therapeutics for human disease. One successful example of applying such approaches is the identification of the LANCL2 pathway as a target of a naturally occurring compound of ABA, which has created a blueprint for a drug development program in autoimmune diseases.

References

Abedi V, Lu P, Hontecillas R, Meghna V, Vess GA et al (2016) Phase III placebo-controlled, randomized clinical trial with synthetic crohn's disease patients to evaluate treatment response. In: Arabnia H, Qe NT (eds) Emerging trends in applications and infrastructures for computational biology, bioinformatics, and systems biology. Elsevier Inc., Cambridge, MA, pp 411–427. https://doi.org/10.1016/B978-0-12-804203-8.00028-6

Afacan NJ, Fjell CD, Hancock RE (2012) A systems biology approach to nutritional immunology–focus on innate immunity. Mol Asp Med 33:14–25

Allen IC et al (2011) NLRX1 protein attenuates inflammatory responses to infection by interfering with the RIG-I-MAVS and TRAF6-NF-κB signaling pathways. Immunity 34:854–865

Allison DB et al (2015) Goals in nutrition science 2015–2020. Front Nut 2:26

Bassaganya-Riera J, Hontecillas R (2006) CLA and n-3 PUFA differentially modulate clinical activity and colonic PPAR-responsive gene expression in a pig model of experimental IBD. Clin Nutr 25:454–465

Bassaganya-Riera J, Hontecillas R (2010) Dietary conjugated linoleic acid and n-3 polyunsaturated fatty acids in inflammatory bowel disease. Curr Opin Clin Nutr Metab Care 13:569–573. https://doi.org/10.1097/MCO.0b013e32833b648e

Bassaganya-Riera J et al (2004) Activation of PPAR γ and δ by conjugated linoleic acid mediates protection from experimental inflammatory bowel disease. Gastroenterology 127:777–791

Bassaganya-Riera J et al (2011a) Activation of PPARγ and δ by dietary punicic acid ameliorates intestinal inflammation in mice. Br J Nutr 106:878–886

Bassaganya-Riera J et al (2011b) Soluble fibers and resistant starch ameliorate disease activity in interleukin-10–deficient mice with inflammatory bowel disease. J Nutr 141:1318–1325

Bassaganya-Riera J, Guri AJ, Hontecillas R (2011c) Treatment of obesity-related complications with novel classes of naturally occurring PPAR agonists. J Obes 2011:897894. https://doi.org/10.1155/2011/897894

Bassaganya-Riera J et al (2011d) Abscisic acid regulates inflammation via ligand-binding domain-independent activation of peroxisome proliferator-activated receptor γ. J Biol Chem 286:2504–2516

Bassaganya-Riera J et al (2012a) Helicobacter pylori colonization ameliorates glucose homeostasis in mice through a PPAR gamma-dependent mechanism. PLoS One 7:e50069. https://doi.org/10.1371/journal.pone.0050069

Bassaganya-Riera J, Hontecillas R, Horne WT, Sandridge M, Herfarth HH, Bloomfeld R, Isaacs KL (2012b) Conjugated linoleic acid modulates immune responses in patients with mild to moderately active Crohn's disease. Clin Nutr 31:721–727

Bassaganya-Riera J et al (2012c) Probiotic bacteria produce conjugated linoleic acid locally in the gut that targets macrophage PPAR γ to suppress colitis. PLoS One 7:e31238

Bassaganya-Riera J, Viladomiu M, Pedragosa M, De Simone C, Hontecillas R (2012d) Immunoregulatory mechanisms underlying prevention of colitis-associated colorectal cancer by probiotic bacteria. PLoS One 7:e34676

Basson A, Trotter A, Rodriguez-Palacios A, Cominelli F (2016) Mucosal interactions between genetics, diet, and microbiome in inflammatory bowel disease. Front Immunol 7:290. https://doi.org/10.3389/fimmu.2016.00290

Blümer N, Pfefferle PI, Renz H (2007) Development of mucosal immune function in the intra-uterine and early postnatal environment. Curr Opin Gastroenterol 23:655–660. https://doi.org/10.1097/MOG.0b013e3282eeb428

Bolick DT et al (2014) Zinc deficiency alters host response and pathogen virulence in a mouse model of enteroaggregative Escherichia coli-induced diarrhea. Gut Microbes 5:618–627. https://doi.org/10.4161/19490976.2014.969642

Bruzzone S et al (2012) The plant hormone abscisic acid increases in human plasma after hyperglycemia and stimulates glucose consumption by adipocytes and myoblasts. FASEB J 26:1251–1260. https://doi.org/10.1096/fj.11-190140

Buffie CG et al (2015) Precision microbiome restoration of bile acid-mediated resistance to Clostridium difficile. Nature 517:205–208. https://doi.org/10.1038/nature13828

Carbo A et al (2013a) Predictive computational modeling of the mucosal immune responses during helicobacter pylori infection. PLoS One 8:e73365

Carbo A et al (2013b) Systems modeling of molecular mechanisms controlling cytokine-driven CD4+ T cell differentiation and phenotype plasticity. PLoS Comput Biol 9:e1003027

Carbo A et al (2014) Systems modeling of the role of interleukin-21 in the maintenance of effector CD4+ T cell responses during chronic Helicobacter pylori infection. MBio 5:e01243–e01214

Chang C-H et al (2013) Posttranscriptional control of T cell effector function by aerobic glycolysis. Cell 153:1239–1251

Cipolletta D et al (2012) PPAR-γ is a major driver of the accumulation and phenotype of adipose tissue Treg cells. Nature 486:549–553

Evans NP, Misyak SA, Schmelz EM, Guri AJ, Hontecillas R, Bassaganya-Riera J (2010) Conjugated linoleic acid ameliorates inflammation-induced colorectal cancer in mice through activation of PPARγ. J Nutr 140:515–521

Everts B et al (2014) TLR-driven early glycolytic reprogramming via the kinases TBK1-IKK [epsiv] supports the anabolic demands of dendritic cell activation. Nat Immunol 15:323–332

Field CJ, Thomson CA, Van Aerde JE, Parrott A, Lien E, Clandinin MT (2000) Lower proportion of CD45R0+ cells and deficient interleukin-10 production by formula-fed infants, compared with human-fed, is corrected with supplementation of long-chain polyunsaturated fatty acids. J Pediatr Gastroenterol Nutr 31:291–299

Foster JA, McVey Neufeld KA (2013) Gut-brain axis: how the microbiome influences anxiety and depression. Trends Neurosci 36:305–312. https://doi.org/10.1016/j.tins.2013.01.005

Gardy JL, Lynn DJ, Brinkman FS, Hancock RE (2009) Enabling a systems biology approach to immunology: focus on innate immunity. Trends Immunol 30:249–262

Guri AJ, Hontecillas R, Bassaganya-Riera J (2010a) Abscisic acid ameliorates experimental IBD by downregulating cellular adhesion molecule expression and suppressing immune cell infiltration. Clin Nutr 29:824–831

Guri AJ, Hontecillas R, Bassaganya-Riera J (2010b) Abscisic acid synergizes with rosiglitazone to improve glucose tolerance and down-modulate macrophage accumulation in adipose tissue: possible action of the cAMP/PKA/PPAR γ axis. Clin Nutr 29:646–653

Haxhinasto S, Mathis D, Benoist C (2008) The AKT–mTOR axis regulates de novo differentiation of CD4(+)Foxp3(+) cells. J Exp Med 205:565–574. https://doi.org/10.1084/jem.20071477

Hontecillas R, O'Shea M, Einerhand A, Diguardo M, Bassaganya-Riera J (2009) Activation of PPAR γ and α by punicic acid ameliorates glucose tolerance and suppresses obesity-related inflammation. J Am Coll Nutr 28:184–195

Hontecillas R et al (2011) Immunoregulatory mechanisms of macrophage PPAR-gamma in mice with experimental inflammatory bowel disease. Mucosal Immunol 4:304–313. https://doi.org/10.1038/mi.2010.75

John S, Luben R, Shrestha SS, Welch A, Khaw KT, Hart AR (2010) Dietary n-3 polyunsaturated fatty acids and the aetiology of ulcerative colitis: a UK prospective cohort study. Eur J Gastroenterol Hepatol 22:602–606. https://doi.org/10.1097/MEG.0b013e3283352d05

Johnson AR et al (2016) Metabolic reprogramming through fatty acid transport protein 1 (FATP1) regulates macrophage inflammatory potential and adipose inflammation. Mol Metab 5:506–526. https://doi.org/10.1016/j.molmet.2016.04.005

Kverka M et al (2011) Oral administration of Parabacteroides distasonis antigens attenuates experimental murine colitis through modulation of immunity and microbiota composition. Clin Exp Immunol 163:250–259

Leber A et al (2015) Systems modeling of interactions between mucosal immunity and the gut microbiome during clostridium difficile infection. PLoS One 10:e0134849

Leber A et al (2016) Modeling the role of lanthionine synthetase C-like 2 (LANCL2) in the modulation of immune responses to helicobacter pylori infection. PLoS One 11:e0167440

Leber A, Hontecillas R, Abedi V, Tubau-Juni N, Zoccoli-Rodriguez V, Stewart C, Bassaganya-Riera J (2017a) Modeling new immunoregulatory therapeutics as antimicrobial alternatives for treating Clostridium difficile infection. Artif Intell Med 78:1–13. https://doi.org/10.1016/j.artmed.2017.05.003

Leber A et al (2017b) NLRX1 regulates effector and metabolic functions of CD4+ T cells. J Immunol 198:2260–2268. https://doi.org/10.4049/jimmunol.1601547

Lee YS et al (2013) The fractalkine/CX3CR1 system regulates β cell function and insulin secretion. Cell 153:413–425

Lu P, Bevan DR, Lewis SN, Hontecillas R, Bassaganya-Riera J (2011) Molecular modeling of lanthionine synthetase component C-like protein 2: a potential target for the discovery of novel type 2 diabetes prophylactics and therapeutics. J Mol Model 17:543–553. https://doi.org/10.1007/s00894-010-0748-y

Lu P et al (2012) Computational modeling-based discovery of novel classes of anti-inflammatory drugs that target Lanthionine Synthetase C-like protein 2. PLoS One 7:e34643. https://doi.org/10.1371/journal.pone.0034643

Lu P et al (2015) Modeling-enabled characterization of novel NLRX1 ligands. PLoS One 10:e0145420. https://doi.org/10.1371/journal.pone.0145420

Luyer MD, Greve JW, Hadfoune M, Jacobs JA, Dejong CH, Buurman WA (2005) Nutritional stimulation of cholecystokinin receptors inhibits inflammation via the vagus nerve. J Exp Med 202:1023–1029. https://doi.org/10.1084/jem.20042397

Lyte M (2013) Microbial endocrinology in the microbiome-gut-brain axis: how bacterial production and utilization of neurochemicals influence behavior. PLoS Pathog 9:e1003726. https://doi.org/10.1371/journal.ppat.1003726

McDonald D, Glusman G, Price ND (2016) Personalized nutrition through big data. Nat Biotechnol 34:152–154. https://doi.org/10.1038/nbt.3476

Mei Y et al (2014) ENISI MSM: a novel multi-scale modeling platform for computational immunology. In: 2014 IEEE international conference on Bioinformatics and Biomedicine (BIBM). IEEE, pp 391–396

Michalek RD et al (2011) Cutting edge: distinct glycolytic and lipid oxidative metabolic programs are essential for effector and regulatory CD4+ T cell subsets. J Immunol 186:3299–3303

Neuman MG (2007) Immune dysfunction in inflammatory bowel disease. Transl Res 149:173–186. https://doi.org/10.1016/j.trsl.2006.11.009

Neurath MF (2014) Cytokines in inflammatory bowel disease. Nat Rev Immunol 14:329–342. https://doi.org/10.1038/nri3661

Newton R, Priyadharshini B, Turka LA (2016) Immunometabolism of regulatory T cells. Nat Immunol 17:618–625

O'Neill LAJ, Kishton RJ, Rathmell J (2016) A guide to immunometabolism for immunologists. Nat Rev Immunol 16:553–565. https://doi.org/10.1038/nri.2016.70

Penders J et al (2006) Factors influencing the composition of the intestinal microbiota in early infancy. Pediatrics 118:511–521. https://doi.org/10.1542/peds.2005-2824

Philipson CW, Bassaganya-Riera J, Viladomiu M, Pedragosa M, Guerrant RL, Roche JK, Hontecillas R (2013) The role of peroxisome proliferator-activated receptor gamma in immune responses to enteroaggregative Escherichia Coli infection. PLoS One 8:e57812. https://doi.org/10.1371/journal.pone.0057812

Philipson CW et al (2015) Modeling the regulatory mechanisms by which NLRX1 modulates innate immune responses to Helicobacter pylori infection. PLoS One 10:e0137839. https://doi.org/10.1371/journal.pone.0137839

Picard F, Auwerx J (2002) PPAR(gamma) and glucose homeostasis. Annu Rev Nutr 22:167–197. https://doi.org/10.1146/annurev.nutr.22.010402.102808

Redig AJ, Jänne PA (2015) Basket trials and the evolution of clinical trial design in an era of genomic medicine. J Clin Oncol 33:975–977

Richard C, Lewis ED, Goruk S, Field CJ (2016) A dietary supply of docosahexaenoic acid early in life is essential for immune development and the establishment of oral tolerance in female rat offspring. J Nutr 146:2398–2406. https://doi.org/10.3945/jn.116.237149

Ruemmele FM (2016) Role of diet in inflammatory bowel disease. Ann Nutr Metab 68(Suppl 1):33–41. https://doi.org/10.1159/000445392

Schuhmacher A, Gassmann O, Hinder M (2016) Changing R&D models in research-based pharmaceutical companies. J Transl Med 14(1):105

Shah R et al (2015) Metabolic effects of CX3CR1 deficiency in diet-induced obese mice. PLoS One 10:e0138317

Singh UP et al (2014) The emerging role of leptin antagonist as potential therapeutic option for inflammatory bowel disease. Int Rev Immunol 33:23–33. https://doi.org/10.3109/08830185.2013.809071

Smith PM et al (2013) The microbial metabolites, short-chain fatty acids, regulate colonic Treg cell homeostasis. Science 341:569–573. https://doi.org/10.1126/science.1241165

Sturla L et al (2009) LANCL2 is necessary for abscisic acid binding and signaling in human granulocytes and in rat insulinoma cells. J Biol Chem 284:28045–28057. https://doi.org/10.1074/jbc.M109.035329

Verma M et al (2016) Modeling-enabled systems nutritional immunology. Front Nutr 3:5. https://doi.org/10.3389/fnut.2016.00005

Verma M et al (2017) Modeling the mechanisms by which HIV-associated immunosuppression influences HPV persistence at the oral mucosa. PLoS One 12:e0168133. https://doi.org/10.1371/journal.pone.0168133

Vodovotz Y et al (2016) Solving Immunology? Trends Immunol 38(2):116–127. https://doi.org/10.1016/j.it.2016.11.006

Vroegrijk IO et al (2011) Pomegranate seed oil, a rich source of punicic acid, prevents diet-induced obesity and insulin resistance in mice. Food Chem Toxicol 49:1426–1430. https://doi.org/10.1016/j.fct.2011.03.037

Wendelsdorf K, Bassaganya-Riera J, Hontecillas R, Eubank S (2010) Model of colonic inflammation: immune modulatory mechanisms in inflammatory bowel disease. J Theor Biol 264:1225–1239

Wendelsdorf KV et al (2012) ENteric Immunity SImulator: a tool for in silico study of gastroenteric infections. IEEE Trans Nanobioscience 11:273–288

Wooden B, Goossens N, Hoshida Y, Friedman SL (2017) Using big data to discover diagnostics and therapeutics for gastrointestinal and liver diseases. Gastroenterology 152:53–67000. https://doi.org/10.1053/j.gastro.2016.09.065

Xu Y et al (2016) Glycolysis determines dichotomous regulation of T cell subsets in hypoxia. J Clin Invest 126:2678–2688

Yurist-Doutsch S, Arrieta MC, Tupin A, Valdez Y, Antunes LC, Yen R, Finlay BB (2016) Nutrient deprivation affects salmonella invasion and its interaction with the gastrointestinal microbiota. PLoS One 11:e0159676. https://doi.org/10.1371/journal.pone.0159676

Zeevi D et al (2015) Personalized nutrition by prediction of glycemic responses. Cell 163:1079–1094. https://doi.org/10.1016/j.cell.2015.11.001

Zeng M, van der Donk WA, Chen J (2014) Lanthionine synthetase C-like protein 2 (LanCL2) is a novel regulator of Akt. Mol Biol Cell 25:3954–3961. https://doi.org/10.1091/mbc.E14-01-0004

Zhang YZ, Li YY (2014) Inflammatory bowel disease: pathogenesis. World J Gastroenterol 20:91–99. https://doi.org/10.3748/wjg.v20.i1.91

Zhou YJ, Gao J, Yang HM, Yuan XL, Chen TX, He ZJ (2010) The role of the lactadherin in promoting intestinal DCs development in vivo and vitro. Clin Dev Immunol 2010:357541. https://doi.org/10.1155/2010/357541

Zocchi E et al (2017) Abscisic acid: a novel nutraceutical for glycemic control. Front Nutr 4:24. https://doi.org/10.3389/fnut.2017.00024

Chapter 5
Development of Synthetic Patient Populations and *In Silico* Clinical Trials

Ramin Zand, Vida Abedi, Raquel Hontecillas, Pinyi Lu, Nariman Noorbakhsh-Sabet, Meghna Verma, Andrew Leber, Nuria Tubau-Juni, and Josep Bassaganya-Riera

Abstract Drug development, which includes clinical trials, is a lengthy and expensive process that could significantly benefit from predictive modeling and *in silico* testing. Additionally, current treatments were designed based on the average patient using the "one size fits all" protocol. Therefore, they can be effective on some patients but not for others. There is an urgent need to replace such generalized approaches with personalized and predictive strategies that capture and analyze

R. Zand
Department of Neurology, University of Tennessee Health Science Center,
Memphis, TN, USA

Department of Neurology, Geisinger Medical Center, Danville, PA, USA

Nutritional Immunology and Molecular Medicine Laboratory, Biocomplexity Institute
of Virginia Tech, Blacksburg, VA, USA

V. Abedi (✉)
Nutritional Immunology and Molecular Medicine Laboratory, Biocomplexity Institute
of Virginia Tech, Blacksburg, VA, USA

Biomedical and Translational Informatics Institute, Geisinger Health System,
Danville, PA, USA
e-mail: vabedi@geisinger.edu

R. Hontecillas · J. Bassaganya-Riera
Nutritional Immunology and Molecular Medicine Laboratory, Biocomplexity Institute
of Virginia Tech, Blacksburg, VA, USA

Biotherapeutics Inc., Blacksburg, VA, USA

P. Lu · A. Leber
Biotherapeutics Inc., Blacksburg, VA, USA

N. Noorbakhsh-Sabet
Department of Neurology, University of Tennessee Health Science Center,
Memphis, TN, USA

M. Verma · N. Tubau-Juni
Nutritional Immunology and Molecular Medicine Laboratory, Biocomplexity Institute
of Virginia Tech, Blacksburg, VA, USA

© Springer International Publishing AG, part of Springer Nature 2018
J. Bassaganya-Riera (ed.), *Accelerated Path to Cures*,
https://doi.org/10.1007/978-3-319-73238-1_5

57

human diversity and variation at a resolution sufficient to identify and clinically validate personalized treatment paradigms. Utilization of heterogenous datasets, such as Electronic Health Records (EHRs), to build synthetic populations of patients and personalized, predictive models of response to therapy holds enormous promise in precipitating a revolution in precision medicine for IBD. *In silico* trials can be designed to include multi-modal data sources, including clinical trial data at the individual and aggregated levels, pre-clinical data from animal studies, as well as data from EHR. *In silico* clinical trials can help inform the design of clinical trials and make prediction at the population and individual level to increase the chances of success. This chapter discusses pioneering work on the use of *in silico* clinical trials to accelerate the development of new drugs.

Keywords Synthetic patient population · *In silico* clinical trials · Artificial intelligence · Machine learning · EHR · IBD

Overview

Drug development is a lengthy and expensive process that involves target validation, pre-clinical testing, investigational new drug (IND) approval, as well as safety and efficacy testing in Phase I and Phase II clinical trials until the new drug application (NDA). Reductionist and traditional approaches hinder the attempt to address the delay in current translational research thereby necessitating a re-evaluation of the clinical trial design (An et al. 2011). For the most part, the current design of the clinical trials does not account for differences present in a heterogeneous population which includes differences in individual expression profile due to age and genetics factors (An et al. 2011), or differences in response rate amongst the patients (Lund et al. 2014; van Die et al. 2009). Thus, there is an unmet clinical need to design safer and more effective therapeutics to re-invent the design of clinical studies and overcome the challenges of drug discovery and development process. Recently, *in silico* approaches have been used in drug discovery, delivery, and development pathway. There are also studies (Abedi et al. 2016; An 2004; Clermont et al. 2004; Kumar et al. 2008; Li et al. 2008) that have described the application of *in silico* clinical trials to suggest better patient selection criteria. The *in silico* approaches involve the use of data-driven bottom-up computational algorithms in combination with top-down mechanistic models to increase the predictive power of drug development. A recent study (Wise and Bar-Joseph 2014) designed a scalable method that integrated the static and time series data from multiple individuals and reconstructed the condition-specific response rate with an aim to analyze the human response to influenza and mouse brain development. A limited number of studies that successfully implemented the use of *in silico* experimentation techniques includes the development of clinical trial simulation tools used in the design of future clinical trials in Alzheimer's disease (Romero et al. 2015). From mathematical models to advanced

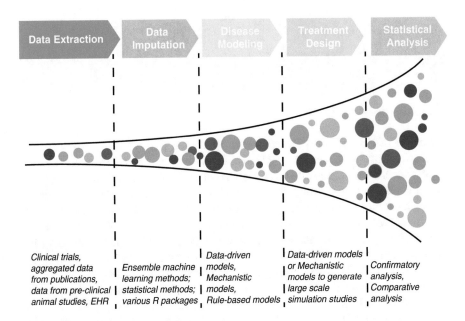

Fig. 5.1 Workflow of *in silico* clinical trial

machine learning methods, the design of *in silico* clinical trials can be performed to optimize clinical trials and alleviate, to some extent, some of the challenges of clinical development plans, including inadequate dosing, inappropriate patient selection, lack of consideration of individual heterogeneity and insufficient design optimization. Overall the workflow of an *in silico* clinical trial can be summarized by five main stages (Fig. 5.1): data extraction, data imputation (if there are missing values), disease modeling, treatment design, and statistical analysis. This chapter will attempt to give an insight into these stages.

Clinical Research and Clinical Trials

Clinical research is a branch of healthcare science that experimentally describes the safety and efficacy of medications, diagnostic or treatment devices for subject use. Clinical research that involves biomedical or behavioral interventions that contribute to the marketing of a treatment are termed clinical trials. Clinical research goes back to ancient history. The Persian physician and philosopher Avicenna (980–1037), in his encyclopedic Canon of Medicine (1025), set down seven rules for evaluation of the effect of drugs on diseases. He suggested that a treatment should be used in its natural state, with uncomplicated diseases, and should be observed in two equal types of diseases. He also indicated that the time of action and duration of the treatment effect should be studied (Meinert 2012). The modern development of clinical research probably started with Sir Ronald A. Fishier. Dr. Fisher developed his

Fig. 5.2 The phases of clinical trial

Principles of Experimental Design in the 1920s and suggested four strategies for decreasing study bias, including random assignment of participants to distinct groups for the experiment (Creswell 2002); replication to alleviate uncertainty, blocking of experimental units into groups that are equivalent, and factorial design so multiple independent factors and possible interactions could be compared (Meinert 2012).

Clinical research can be classified as two groups of products examined, devices and medications, as well as research on techniques. Most frequently clinical trials evaluate the safety and efficacy of a newly developed medication, and the trials seek to determine if the benefits outweigh the risks (FDA 2015; PhRMA 2015).

In the United States, clinical trials involving new drugs are commonly classified into four phases before approved by the national regulatory authority for use in the general population. Each phase (Fig. 5.2) of the drug approval process is treated as a separate clinical trial (Friedman et al. 2010):

- Preclinical IND-enabling phase: Testing of drug in non-human subjects, to gather efficacy, toxicity and pharmacokinetic information. Dose is unrestricted.
- Phase I: Testing of drug on 20–100 healthy volunteers for dose ranging and medication safety, often starting with sub-therapeutic doses and including ascending doses. It usually includes pharmacokinetic and pharmacodynamic (PK/PD). The number of subject will also depend on the disease population.
- Phase II: Establishing the efficacy of the drug, usually against a placebo. Includes 80–300 patients, depending on the disease population, on a proposed therapeutic dose.
- Phase III: Final confirmation of safety and efficacy on large populations of patients using a proposed therapeutic dose and formulation. Side effects are monitored, and the new therapeutic is compared to commonly used treatments. Information is collected that will allow the intervention to be used safely.
- Phase IV: Post marketing surveillance of the marketed drug to evaluate long-term safety and effects.

Design Strategies

Clinical study design is the formulation of trials and experiments, as well as observational studies in medical, clinical or epidemiological research involving human. The goal of a clinical study is to assess the safety, efficacy, or the mechanism of action of an investigational product or procedure, but potentially not yet approved by a health authority like FDA in the United States (Porta 2014).

A fundamental distinction in evidence-based practice is divided in two categories of observational studies (cross sectional, cohort or case control studies) or randomized controlled trials that include phase 2 and 3 drug trials. They are often double blind and placebo controlled. Although the term "clinical trials" is most commonly associated with the large, randomized studies typical of phase 3, many clinical trials are small to test simple questions.

Clinical trial design has its roots in classical experimental design, yet has some different features. The clinical investigators are not able to control as many sources of variability through design as a laboratory or industrial investigators can. Human responses to medical treatments display greater variability than genetically identical plants and animals. Ethical issues are paramount in clinical research. Subject enrollment can become a lengthy and complicated process. In addition, to study a clinical response with adequate precision, a trial requires lengthy periods for patient accrual and follow-up that can be very costly. These are the driving force for the development of *in silico* clinical trials to facilitate design of clinical trials that can be more successful.

Data Sources

Building models that span different spatiotemporal scales and capture diverse set of modalities require data from various sources. For instance, clinical trial data at the individual and aggregated levels can provide valuable insights into building large *in silico* studies. Pre-clinical data from animal studies also provide a unique opportunity to extrapolate from animal model studies. Electronic Health Records (EHR) can also be a valuable resource.

EHR can provide information at different levels (i.e. social status, behavioral data, lab measure, even zip codes in some instances); however, this type of data present unique challenges, including missing value among others. In general terms, missing data in EHR can be classified as (1) Incomplete, (2) Inconsistent, and (3) Inaccurate. Incompleteness relates to Inherent missing data due to ascertainment bias; for instance, patients may present at irregular intervals and can move between health care providers. Inconsistency relates to the process in which data collection and storage vary by location or provider practice. Inaccuracy occurs by varying levels of specifics, standards, and precision that can exist in the care process, causing inexact or incorrect information. In other words, missing data can be due to the fact that outcomes and diagnosis may not have been recorded in a timely fashion or may be missing from the patient chart. For example, the patient could have been diagnosed and cared for outside of the given health care system, did not seek treatment, or the health care provider did not enter the information. It is also possible that the patient has died and the information is not updated in the system. Missing data are an important challenge when utilizing EHR data for modeling purposes. Understanding the kind of missing information is key in the modeling process. In particular, there are three main categories: missing-completely-at-random (MCAR), missing-at-random (MAR), and not-missing-at-random (NMAR). A comprehensive overview of missing data in EMR is presented in (Wells et al. 2013). To address this challenge, it is always possible to omit the cases with missing data and run the analysis on the full dataset; however, that causes bias in the analysis and significantly limits the power of the study by limiting the sample size. There are methods such as "multiple imputation using chained equations" (MICE) (White et al. 2011) that are based on approximating the posterior predicted distribution of each variable by regressing it on all other remaining variables. This method can impute different variable types. Nonetheless, from a statistical point, it will be a question on how many imputed data sets will be required to obtain a stable estimate. Furthermore, the sample size is also a critical component. When the number of subjects is limited, the training set is small, and it will be more challenging to find a stable estimate to represent the missing values. Alternatively, different machine learning methods can be used to obtain a different estimate for the missing values based on the available data on the subjects. Alternatively, if the number of cases is large, it is possible to use more targeted strategy for imputation. In a recent study of 11K patients, a novel technique using latent-based analysis integrated with clustering was used to group patients based on their comorbidities before imputation. The imputation approach had lower median root-mean-square error (RMSE) gain when compared to standard methods.

Using EHR, it is also possible to build predictive personalized models of diseases and response to treatments. Integrating EHR data with other sources of clinical and pre-clinical data can provide valuable opportunities for knowledge discovery by providing venues for building models that can generate non-intuitive hypotheses. Finally, genetic data that are linked to EHR can be very effective in driving the path to personalized medicine and health. For instance, a partnership between Geisinger, pioneering in digital medical records, and Regeneron, a leading science-based bio-pharmaceutical company, is such example; this unique partnership has the potential to transform medicine and, to predict and anticipate diseases before the onset and optimize health at many levels. These types of large-scale collaborations are unique assets and provide a tremendous opportunity for advancement of precision medicine initiative (Abul-Husn et al. 2016; Dewey et al. 2016) as well as new opportunities for modeling-enabled and data-driven discoveries.

Finally access to data, especially raw data that can be re-analyzed or used for calibration of internal components of computational models, can lead to more success in predictive analysis and hypothesis generation and to help in developing *in silico* clinical trials to fine tune and optimize design of clinical trials. An example of such data includes but not limited to data from clinical trials at a patient level, de-identified EHR, as well as pre-clinical data at different scales (from the cellular level to tissue level).

Predictive Modeling for Precision Medicine

Scientists ranging from physicians to molecular biologists, mathematicians, physicists, and computer scientists are working together to solve the problem of understanding and curing diseases. In the process, predictive modeling approaches or tools play a key role in designing effective vaccines and drugs to reverse or slow the progression of the disease at the molecular level. However, control of disease progression and prevention is impeded by multiple factors ranging from host-pathogen to host-host interactions, or social, economical and demographic conditions that affect the progression and manifestation of a disease (Siettos and Russo 2013). With the advent of complex multi-scale techniques, there are an array of mathematical and statistical modeling strategies, tools and techniques that are aimed towards understanding the spread of the disease (Janes and Yaffe 2006). These methods aid in the prediction of the behavior of the system to gain deeper insights on the underlying complex interplay among different contributing elements, such as genes, pathways, microbiome, nutrition, or environmental factors.

Data-Driven Models

Data-driven models are extensively used, especially with growing computational resources and availability of large-scale clinical datasets. They have been shown to be instrumental for analyzing mechanistic biological data (Janes and Yaffe 2006). The approach includes the use of *-omics* data ranging from lipidomics, metabolomics, epigenomics and proteomics that give an insight regarding a biological condition when the model is trained on certain disease-specific conditions. These models help users analyze large clinical datasets by simplified measurements with a minimal set of assumptions and constraints. Since the models are based on the analysis of available data, the performance is determined by the access to the data. One of the main limitations of data-driven modeling approach includes the lack of integrated mechanistic knowledge of the biological system, and the inability to draw conclusions based on that (Janes and Yaffe 2006). The most common data-driven modeling approach that are based on statistical procedures include for example Exploratory Data Analysis (EDA), Principal component analysis (PCA), Linear discriminant analysis (LDA), Latent Semantic Analysis (LSA), Factor Analysis, Cluster Analysis (which can be based on hard clustering, soft clustering or hierarchical clustering), Regression-based methods (such as Partial Least Squares), and various Machine Learning methods (such as Artificial Neural Networks, Random Forest or, Support Vector Machine (SVM)). However, even though these methods do not consider prior knowledge, they allow integration of additional constraints to ensure the results are aligned with biologically coherent concepts or independent clinical information.

There are different ways that data-driven modeling can help extract quantitative information and knowledge from large datasets. For instance, clustering techniques can aid in the reduction of the dimensionality of the data which reveals useful information such as the related signal transduction proteins, gene networks, and biological pathways. Thus, by reducing dimensionality, concepts, genes, or biological entities that are closely related in terms of function can be visualized in a reduced dimension. For example, PCA is a statistical procedure that results in the reduction of the dimensionality of the data by generating linearly uncorrelated variables termed as principal components (PC), which capture the variance in the data. The PCs allow users to visualize a large amount of data that contain a set of variables that are correlated. The components condense the data such that they are always less than or equal to a number of variables in the data set. The PCA provides key information regarding the axes that contain the most important information required to differentiate the data variables. The underlying methodology in PCA involved in creating the PCs comprises of mathematical approaches that scale and normalize the data. These components thereby highlight the important features that capture the maximum covariance between the variables (Janes and Yaffe 2006). Thus, PCA is a method of data organization that condenses the data and makes the visualization simple by projecting it on the PCs of the dataset under consideration. The partial least squares (PLS) is another technique, where data is divided into independent and dependent variables and a linear solution which relates these variables is determined. The major difference between the two is that, in PCA, the conclusion is based on the original data space whereas PLS reduces the data to key principal components with the prediction of independent components from dependent components. Another difference is that the loading vectors in PLS capture optimal covariance between the dependent and independent components, in spite of the reduced efficiency in capturing the data when compared to PCA (Janes and Yaffe 2006).

Applications of exploratory analysis, through use of dimensionaliry reduction as well as advanced machine learning of large datasets include the use of (1) PLS to analyze tumor classification based on the microarray gene expression data (Nguyen and Rocke 2002), (2) modeling and simulation of the signal transduction cascades and genetic regulatory networks of cellular behavior (Newell et al. 2012), (3) modeling and classification (including machine learning methods) of different cell types (Prilutsky et al. 2011), machine learning in recognition of acute cerebral ischemia (ACI) and differentiating that from stroke mimics in an emergency setting (Abedi et al. 2017) among others. Large scale data-driven studies can help in the design of more successful clinical research, including clinical trials, by for instance improving patient stratification, treatment duration and inclusion/exclusion criteria.

Mechanistic Models

Mechanistic models are mainly based on the literature-derived information and knowledge from the experts in the fields. These include equation-based models such as ordinary differential equations (ODEs), partial differential equations (PDEs) and stochastic differential equation (SDEs). The modeling exercise can also involve simulation of key elements, including intracellular, intercellular, tissue, organ system and interactions that connect all the different levels of modeling termed as multiscale modeling. These mechanistic models can also be integrated with agent-based models (ABMs) for more complex systems. This modeling framework represents the dynamics of the system under investigation and can be derived from physiological interactions which are determined by principles of mass action kinetics. Such models are extremely powerful due to the consideration of the underlying biological mechanisms. However, one of the main drawbacks includes the unintended bias towards a pathway that involves a set of the network of proteins or gene-gene interaction. In addition, developing such models requires extensive literature search and close collaboration with of domain experts for interpreting the data and the findings in addition to setting suitable constraints.

The system of ODEs is the most widely used approach for mechanistic modeling. The variables in an ODE model are represented using continuous time, such that the system of equations simulates the concentration of each variable over time. The ODE-based approach is extensively used to study numerous biological pathways (Leber et al. 2016; Machado et al. 2011), response against infectious agents (Carbo et al. 2013), heterogeneity of the immune cell subsets and the body's response against inflammatory bowel disease (Carbo et al. 2014; Hoops et al. 2015; Wendelsdorf et al. 2010). The equations are used to investigate time course simulations, steady state of the behavior of the system and possibility of bifurcation points in the system that drives the system from a stable to an unstable state. The main advantage of ODE-based models lies in information regarding the variation of species over time. However, one of the drawbacks of ODEs includes the requirement of prior reaction mechanisms to ensure the proper use of rate laws. Another hurdle lies in the availability of experimental data that needs to be utilized for model calibration. The outcome from these models can be valuable in integrative analysis and time dependent predictions.

Rule-Based Models

The mechanistic models also comprise of *rule-based models*. The "rules" are defined for the transformation of the states of the components, in contrast to defining one rule for each state (Machado et al. 2011). The interactions within the system can be visualized using contact maps. The main advantage of the rule-based approach is the stochasticity in the simulations which generates the species and create reactions as they are available (Carbo et al. 2013). *Agent-based models* are similar to rule-based models wherein each element of the system is referred to as an agent. The agent has an increased capability to interact with another agent, and they move free within the spaces. The agent behavior in the system is governed by certain rules which can help in the understanding of the complex phenomenon. These models have been majorly used to take into account the spatial distribution of the system. Their major application is involved in the use of the multi-cellular level; wherein they have been used to study the immune response to an infectious agent such as *Helicobacter pylori* (Carbo et al. 2013), tumor growth (Carbo et al. 2013; Zhang et al. 2007), granuloma formation (Segovia-Juarez et al. 2004) and chemotaxis. Thus, in comparison to the equation-based models that predict the concentrations over time, ABMs comprise of each individual model entity that has the ability to interact with other agents in their proximity and are useful to study the effect of local interactions on the large scale behavior (Forrest and Beauchemin 2007). The main disadvantage of the rule-based and agent-based model include computational intensive calculations and costs involved in increasing the scalability of the system. Furthermore, it is challenging to examine a rule or agent-based system since the output is stochastic in nature.

Human Avatars: Virtual Patient Populations

Virtual patients are key components in the design of *in silico* clinical trials. A virtual or synthetic population carries the attributes of the original patient population and reflects the individual diversity, variation and other individual characteristics that represent a clinical cohort population. The virtual population can be extremely useful in identifying the broad range of responses that are otherwise hard to predict in a clinical trial study. Here we briefly describe five different studies designing virtual population for clinical application.

1. A study by Schmidt et al. described the MAPEL (Mechanistic Axes Population Ensemble Linkage) algorithm to develop a group of calibrated virtual population to enables the development and enhance the interpretation of simulation results (Schmidt et al. 2013a). The algorithm links the output from virtual patients to the clinical outcomes at the population scale. The algorithm utilizes data within the mechanistic parameters underlying the simulations. The set of virtual patients first developed varied regarding the mechanistic axes coefficients (Schmidt et al. 2013a). The variability on the pathophysiological axes was maintained within the cohort of virtual patients by introducing variation in mechanistic axes that were chosen on the bases of sensitivity and diversity. The chosen virtual patients were calibrated manually such that the model simulation results matched closely to the observed clinical responses. A group of additional virtual patients derived from the 'seed' virtual patients was created using a genetic algorithm on a cluster. The range of the parameters varied in the virtual population was obtained from the literature and used for validation of the cohort. To create a virtual population, the MAPEL algorithm assigns a prevalence weight to each virtual patient, which is a value that represents the fraction of that patient present in the virtual population. The prevalence weight is determined by assigning a probability distribution on each of the mechanistic axes for every virtual patient (Schmidt et al. 2013a). Based on the assigned prevalence weights, the created virtual population created is defined by the weights for the virtual patients. Thus, the method involves weighing of the model components for creation of the virtual population.

2. The method developed by Allen et al. generates virtual patients with all the virtual patients weighted equally thereby avoiding the problem of overweighting certain axes (Allen et al. 2016). The approach includes generation of the 'seed' virtual patients which represent a parameter set for each component of the model

(Allen et al. 2016). An initial parameter guess is performed, and the choices are optimized to calibrate the output with the outputs obtained in a clinical setting. The probability of inclusion is calculated for each virtual patient, and the population is selected based on how closely it matches the features of the population of interest. The method describes the importance of a balanced representation of a 'seed' population which avoids bias against certain parameters or axes.

3. The method described by Haidar et al. builds a stochastic virtual population through a process termed 'stochastic cloning' (Haidar et al. 2013). A time-varying model of glucose regulation along with information derived using Bayesian inference were used to estimate the individual parameters of the virtual population. The process of creating a virtual population from the patient data involves the prediction of model parameters which represent specific patient population. This combines both the measured and prior information about the parameters derived using Bayesian approaches.

4. In another recent study, virtual patients were designed based on a multiscale approach. The method (Abedi et al. 2016) involves utilization of clinical trial data (individual response to treatment) integrated with population level data from published clinical trials (aggregated data), expert knowledge, and data from animal studies to create a diverse set of 10,000 virtual patients with Crohn's disease. The goal of the study was to compare current and investigational treatment regimens. The virtual patients were designed to provide diversity and were treated with virtual treatment strategies based on changes in immunological parameters that drive response to treatment.

5. Virtual patients can also be designed based on mechanistic modeling. In an *in silico* clinical trial study of trauma, virtual patients were designed based on a multicompartment ODE model (Brown et al. 2015). The mathematical model was used to create both individual-specific variants and a large cohort of virtual patients to predict trauma-induced mortality and a novel role for interleukin-6.

In summary, in an era of Big Data, limited resources and high costs for preclinical and clinical experimentation and extensive regulatory procedures, it is vital to utilize advanced data analytics and predictive modeling to best optimize the resource allocation and improve the translational impact of large studies. The complexity of clinical trials requires multi-level and multi-modal strategies that can be achieved by integrating data, meta-data, and theory at different spatiotemporal scales. It is also imperative to utilize virtual patients for *in silico* clinical trials to improve this complex process.

Treatment Design

Drug discovery and development of useful therapeutic agents are complex, expensive, and time-consuming, with low success rates in for products proceeding through clinical trials. The research and development (R&D) cost of a new drug was estimated to be \$2.6 billion (pre-approval cost) and 2.9 billion (including post-approval cost) by a recent study by the Tufts Center for the Study of Drug Development. The rising costs are mainly led by increased sizes of clinical trial and higher failure rates for drugs (Phippard 2015). Therefore, there is an urgent need to improve cost-to-value in drug development. To meet this need, the researchers and developer should only invest in those compounds that are likely to succeed in clinical trials. Alternatively, it is better to have failure of an unsuccessful investigation early in preclinical or Phase I studies instead of during expensive Phase III clinical trials (Schmidt et al. 2013b). Testing treatments through predictive *in silico* clinical trials could significantly improve the efficiency and reduce costs during the drug development process. Finally, for designing treatments *in silico*, two useful approaches can be applied: data-driven and mechanistic-based approaches. Data-driven approaches build the predictive models by inferring interactions of molecules without *a priori* knowledge; while, mechanistic-based approaches incorporate information from prior knowledge and simulate models by solving a series of mathematical equations (Clegg and Mac Gabhann 2015).

Data-Driven Design

Data-driven design approaches collect and utilize multiple types of historical data throughout the whole drug development process. The data sources include preclinical studies, clinical trials, and post-marketing surveillance. The data include pharmacokinetics, pharmacodynamics, genomic, electronic health records, and more. After data processing, these data can be used to build models that predict the efficacy and safety of virtual treatments. Frequently applied data-driven approaches include transfer function models, network-centric models, autoregressive time series analysis, nonlinear time series analysis, and Voltera integral series analysis (Aerts et al. 2014).

One of feature of biomedical data is its volume. For instance, the current listing of biomedical research-related information currently spans 1685 databases (Rigden et al. 2016). NCBI's PubChem has more than 157 million chemical substance descriptions with 60 million unique chemical structures and over one million biological assay descriptions, as well as ten million unique protein target sequences (Kim et al. 2016). DrugBank database (Version 5.0) contains 8250 drug entries that cover 2016 FDA-approved small molecule drugs (Law et al. 2014).

Medical data require effective analytics. Machine learning (ML) methods can be used to learn patterns in data and use these patterns for future predictions. ML methods can discover the complex signals underlying biological response to drugs and thus they have been widely used in studies on human biology and drug response (Gertrudes et al. 2012). Supervised machine learning methods use training data to learn the structure of a system and utilize that knowledge to predict the outcome for an unseen condition. Also, they can be used to reduce a large data set to a more manageable set of biomarker candidates (Lu et al. 2015). Artificial neural network and random forest are two examples of supervised machine learning methods. Artificial neural networks algorithms are inspired by the biological neural systems, which are powerful in modeling and data mining tools based on the theory of connectionism. Random forest method is based on the aggregation of a large number of decision trees (Lu et al. 2015). Stanford and Google developed massively multitask networks for drug discovery, which incorporated deep learning and multitask networks. They collected 37.8 million measurements for 1.6 million compounds across more than 200 biological targets to train the model and showed that their multitask networks could generate significantly more accurate predictions than single-task methods (Ramsundar et al. 2015).

Data-driven approaches usually generate models with compact structures, which describe the dominant pattern in the data. These methods have also some limitations; for instance, they highly rely on the availability and quality of sampled data, furthermore, they usually only collect input and output data from a process and thus they cannot provide insight related to the internal state of the process (Aerts et al. 2014).

Mechanistic Modeling-Based Design

The most common reasons for drug failure in clinical trials are inefficiency and toxicity. Both occur due to lack of understanding of mechanisms of action and potential drug targets. This is also the reasons why there are only 5–6 first-in-class drugs approved by the FDA per year (Clegg and Mac Gabhann 2015). Thus, it is important to design drugs based on prior knowledge about mechanisms of drugs and diseases. Mechanistic modeling-based approaches simulate interactions between entities and the processes they undergo (Clegg and Mac Gabhann 2015). This can be used for earlier identification of ineffective or toxic drugs.

There are two main types of mechanistic modeling-based approaches that have been used in drug research and development. The first type of approach builds the disease phenotype-driven models by integrating prior knowledge on molecules, cells, tissues, and biological processes relevant to disease. Phenotype-driven models are usually composed of a set of ordinary differential equations (ODEs). Parameters of models are estimated using literature or experimental measures. Results from model simulations are expected to predict dynamics of therapeutic drug levels, mediator concentrations, cell composition, and tissue-level function. The second mechanistic modeling-based approach constructs the entire network instead of focusing on a specific disease phenotype. The impact of pathologic or therapeutic alterations on a disease network can be predicted by simulating and analyzing activities of a network in situations that include information on disease and health status. Several networks have been applied in drug discovery, such as interaction networks, regulatory networks, signaling networks, and metabolic networks (Schmidt et al. 2013b).

Mechanistic modeling-based approaches can provide insights into mechanisms of action of drugs. In addition, mechanistic modeling-based approaches can develop hypotheses about the characteristics of a disease system. They can also integrate data from multiple sources, which facilitates translating a result from in vitro or animal studies to better predict efficacy in clinical trials. Limitations of mechanistic modeling-based approaches include the need for prior knowledge and experimental information that are not completely available in most cases.

Statistical Analysis

Clinical studies involve investigating the efficacy and safety of proposed medical treatments, assessing the relative benefits of competing therapies, and establishing optimal treatments or treatment combinations. Statistics are an important part of inferential processes (Piantadosi 2013). They provide formal accounting for sources of variability in patients' responses to treatment and allow the clinical researcher to form reasonable and accurate inferences for making sound decisions in the presence of uncertainty (King et al. 2016).

An *in silico* clinical trial is an experimental method for testing medical treatments on virtual patients *in silico*. Statistics are needed at every stage of *in silico* clinical trials, including planning of the study, exploring the clinical data, creating the virtual patients, simulating the treatment process, and interpreting the results. Two main types of statistical analyses are used in the *in silico* clinical trials, confirmatory analysis and comparative analysis. Confirmatory analysis, such as principal component analysis (PCA), can show which parameters/variables are important in large clinical datasets and whether they correlate with what is expected. Comparative analysis, such as analysis of variance (ANOVA), is used to compare effects of new medications with a placebo or standard treatment.

Dimensionality Reduction

For a large dataset with multiple variables, the dispersion matrix may be too large to interpret properly due to too many pairwise correlations between variables. Thus, it is necessary to reduce the number of variables to interpret the data in a more meaningful form. Principal Component Analysis (PCA) is one of the commonly used statistical procedure that reduces the dimensionality of the data while retaining most of the variation in the dataset by identifying principal components (PCs) (Hotelling 1933).

PCA reduces the dimensionality of the original multivariable dataset by finding a linear combination of those variables that explains most of the variability within the considered dataset. Systematic information that is initially dispersed in a large matrix of input variables can be extracted and condensed to a few abstract variables using PCA. PCs can be identified given either the original variables or a correlation or covariance matrix. Coefficients of each PC are determined by computing the eigenvalues of the covariance matrix or the correlation matrix. The PCA procedure is described as follows (Principal Component Analysis (PCA) Procedure 2016):

Suppose that there is a random vector X with a population variance-covariance matrix, var.(X).

$$X = \begin{pmatrix} X_1 \\ X_2 \\ \vdots \\ X_p \end{pmatrix}$$

$$\mathrm{var}(X) = \Sigma = \begin{pmatrix} \sigma_1^2 & \sigma_{12} & \cdots & \sigma_{1p} \\ \sigma_{21} & \sigma_2^2 & \cdots & \sigma_{2p} \\ \vdots & \vdots & \ddots & \vdots \\ \sigma_{p1} & \sigma_{p2} & \cdots & \sigma_p^2 \end{pmatrix}$$

The first PC, PC1, is the linear combination of x-variables that has maximum variance among all linear combinations, so it accounts for as much variation in the data as possible.

$$PC1 = e_{11}X_1 + e_{12}X_2 + \ldots + e_{1p}X_p$$

Coefficients e_{11}, e_{12}, ..., e_{1p} for PC1 are defined in such a way that their variance is maximized, subject to the constraint that the sum of the squared coefficients is equal to one. This constraint is required so that a unique answer can be obtained.

More formally, select e_{11}, e_{12}, ..., e_{1p} that maximizes

$$\mathrm{var}(PC1) = \sum_{k=1}^{p}\sum_{l=1}^{p} e_{1k}e_{1l}\sigma_{kl} = e_1' \Sigma e_1$$

subject to the constraint that

$$e_1' e_1 = \sum_{j=1}^{p} e_{1j}^2 = 1$$

The i-th PC is the linear combination of x-variables that accounts for as much of the remaining variation as possible, with the constraint that the sums of squared coefficients add up to one along with the additional constraint that this new component will be uncorrelated with all the previously defined components.

Interpretation of the PCs is based on finding which variables are most strongly correlated with each component. Therefore, the correlations between the original data for each variable and each principal component need to be computed. Larger correlations of variables in PCs indicate more important variables. For example, in an *in silico* clinical trial to evaluate a new medication for Crohn's disease, PCA analyses on the clinical data were used to confirm that IFN-γ and TNF-α are molecular markers of Crohn's disease (Abedi et al. 2016).

Even though PCA is the mostly utilized dimensionality reduction technique, there are other well-known methods that can be used to achieve similar goals, including linear discriminant analysis (LDA). LDA is used to find a linear combination of variables (or features) that characterizes or best separates two or more classes of objects. LDA is closely related to analysis of variance (ANOVA) and regression; however, ANOVA uses categorical independent variables and a continuous dependent variable, whereas LDA has continuous independent variables and a categorical dependent variable.

Analysis of Variance

Analysis of Variance (ANOVA) is a test that provides a global assessment of a statistical difference in two or more than two independent means. *In silico* clinical trials usually include more than two comparison groups. In an *in silico* clinical trial to evaluate a new medication for Crohn's disease, we compared an experimental medication to a placebo, conjugated linoleic acid (CLA), and GED-0301 (Abedi et al. 2016). The ANOVA procedure was used to compare the means of the comparison groups and was conducted using a five-step approach: (1) Set up hypotheses and select the level of significance; (2) Select the appropriate test statistic; (3) Set up decision rule; (4) Compute the test statistic; (5) Conclusion. Because there are more than two groups, the test statistic needed to take into account the sample sizes, sample means, and sample standard deviations in each of the comparison groups (Sullivan 2016). ANOVA takes into account the total data by asking a global question, such as whether the means of several groups are equal. The fundamental strategy of ANOVA is to systematically examine variability within groups and among the groups being compared (Sullivan 2016).

When analyzing the effect of treatments (placebo, TNF-a blockers, CLA, GED-0301, and LANCL2 therapeutics) and initial Crohn's Disease Activity Index (CDAI) in synthetic populations of patients with Crohn's disease, on the drop in CDAI. (Abedi et al. 2016) the computational simulations predicted that the effects of GED0301 were not as impressive as anticipated. Interestingly, Phase 2b and 3 clinical trials in Crohn's disease patients treated with GED0301 were halted by Celgene, presumably for not reaching the needed level of mucosal healing.

Concluding Remarks

Developing a new medicine takes an average of 10–15 years, and costs almost three billion with an annual growth rate of 8.5%. The likelihood that a new drug will make it into the market is 1 in over 10,000 (DiMasi et al. 2016). Nearly three-quarters of the cost or drug development are spent in multiple phases of clinical trials, due to increasing the size and decreasing success rate as potential therapeutics pass through the phases of these clinical trials. So far, the only way to ensure the efficacy and safety of medication is to test the medication on humans. The clinical studies often include three phases prior to getting approved and post-marketing studies (Viceconti et al. 2016). In Phase I, a small number of patients or healthy people are recruited to examine safety and pharmacokinetics of potential medications. Phase II studies recruit and enroll a sufficient number of patients to test effectiveness of the medication in treatment of disease. Phase III trials require an even larger patient population. Costs for Phase II and Phase III trials can be considerable. Therefore, ending an unsuccessful investigation early in preclinical or Phase I studies instead of Phase II or Phase III trials in actual patients can substantially reduce the cost of drug development (Schmidt et al. 2013b). *In silico* clinical trials have the potential to predict the safety and efficacy of medications and also reduce the cost of clinical trials by (Viceconti et al. 2016): (1) Reducing the size or the duration of clinical trials by predicting individualized adverse side effects and response (or in some cases identify non-responders in advance); (2) Complementing clinical trials by creating experimental test scenarios that can be rare in real patient cohorts; (3) Refining clinical trials by providing more detailed insights on potential outcomes, mechanisms of any adverse side effects, and interactions between tested medications; and (4) Partially replacing clinical trials with *in silico* clinical trials when *in silico* clinical trial could generate scientifically robust evidence under appropriate conditions in certain cases (Fig. 5.3).

  In silico clinical trials, in combination with modeling strategies and multi-modal data sources, will better inform the drug development process. This holds enormous promise in improving the cost efficiency problem of the lengthy and costly drug development process. Technologies designed to predict how patients with different diseases respond to innovative therapeutics can help transform medicine from an art into a more precise science. While *in silico* clinical trials in virtual patients will not completely replace clinical testing in humans, it will inform the design of such studies to increase the likelihood of success and help address the efficiency problem in drug development. *In silico* clinical trials can also optimize innovation and help address an unmet clinical needs for more individualized therapeutic and prophylactic approaches (Abedi et al. 2016; Leber et al. 2017). Furthermore, the combination of mechanistic modeling with data-driven AI approaches such as advanced machine learning algorithms has the potential to bridge the gap between preclinical findings and clinical outcomes in a more predictive and efficient manner.

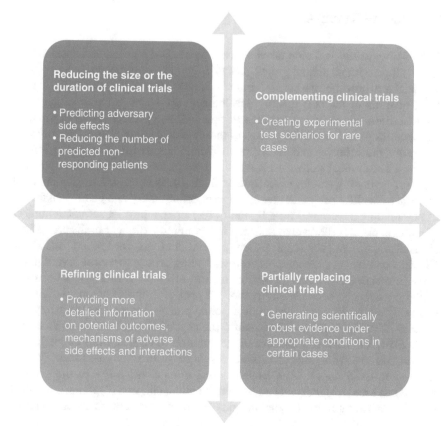

Fig. 5.3 Potentials of *in silico* clinical trials

References

Abedi V et al (2016) Phase III placebo-controlled, randomized clinical trial with synthetic crohn's disease patients to evaluate treatment response. In: Emerging trends in applications and infrastructures for computational biology, bioinformatics, and systems biology. Elsevier, Cambridge, MA, pp 411–427. https://doi.org/10.1016/B978-0-12-804203-8.00028-6

Abedi V et al (2017) Novel screening tool for stroke using artificial neural network. Stroke 48:1678–1681. https://doi.org/10.1161/STROKEAHA.117.017033

Abul-Husn NS et al (2016) Genetic identification of familial hypercholesterolemia within a single U.S. health care system. Science 354:pii: aaf7000. https://doi.org/10.1126/science.aaf7000

Aerts JM, Haddad WM, An G, Vodovotz Y (2014) From data patterns to mechanistic models in acute critical illness. J Crit Care 29:604–610. https://doi.org/10.1016/j.jcrc.2014.03.018

Allen RJ, Rieger TR, Musante CJ (2016) Efficient generation and selection of virtual populations in quantitative systems pharmacology models. CPT Pharmacometrics Syst Pharmacol 5:140–146. https://doi.org/10.1002/psp4.12063

An G (2004) In silico experiments of existing and hypothetical cytokine-directed clinical trials using agent-based modeling. Crit Care Med 32:2050–2060

An G, Bartels J, Vodovotz Y (2011) In silico augmentation of the drug development pipeline: examples from the study of acute inflammation. Drug Dev Res 72:187–200

Brown D et al (2015) Trauma in silico: Individual-specific mathematical models and virtual clinical populations. Sci Transl Med 7:285ra261

Carbo A et al (2013) Predictive computational modeling of the mucosal immune responses during helicobacter pylori infection. PLoS One 8:e73365. https://doi.org/10.1371/journal.pone.0073365

Carbo A, Hontecillas R, Andrew T, Eden K, Mei Y, Hoops S, Bassaganya-Riera J (2014) Computational modeling of heterogeneity and function of CD4+ T cells. Front Cell Dev Biol 2:31. https://doi.org/10.3389/fcell.2014.00031

Clegg LE, Mac Gabhann F (2015) Molecular mechanism matters: benefits of mechanistic computational models for drug development. Pharmacol Res 99:149–154. https://doi.org/10.1016/j.phrs.2015.06.002

Clermont G, Bartels J, Kumar R, Constantine G, Vodovotz Y, Chow C (2004) In silico design of clinical trials: a method coming of age. Crit Care Med 32:2061–2070

Creswell JW (2002) Educational research: planning, conducting, and evaluating quantitative. Prentice Hall, Upper Saddle River, NJ

Dewey FE et al (2016) Distribution and clinical impact of functional variants in 50,726 whole-exome sequences from the DiscovEHR study. Science 354:pii: aaf6814. https://doi.org/10.1126/science.aaf6814

DiMasi JA, Grabowski HG, Hansen RW (2016) Innovation in the pharmaceutical industry: new estimates of R&D costs. J Health Econ 47:20–33. https://doi.org/10.1016/j.jhealeco.2016.01.012

FDA (2015) FDA's drug review process: continued. https://www.fda.gov/Drugs/ResourcesForYou/Consumers/ucm289601.htm. Accessed 18 Oct 2017

Forrest S, Beauchemin C (2007) Computer immunology. Immunol Rev 216:176–197. https://doi.org/10.1111/j.1600-065X.2007.00499.x

Friedman LM, Furberg C, DeMets DL (2010) Fundamentals of clinical trials. Springer, New York

Gertrudes JC, Maltarollo VG, Silva RA, Oliveira PR, Honorio KM, da Silva AB (2012) Machine learning techniques and drug design. Curr Med Chem 19:4289–4297

Haidar A, Wilinska ME, Graveston JA, Hovorka R (2013) Stochastic virtual population of subjects with type 1 diabetes for the assessment of closed-loop glucose controllers. IEEE Trans Biomed Eng 60:3524–3533. https://doi.org/10.1109/TBME.2013.2272736

Hoops S, Hontecillas R, Abedi V, Leber A, Philipson C, Carbo A, Bassaganya-Riera J (2015) Ordinary differential equations (ODEs) based modeling. In: Computational immunology: models and tools, p 63

Hotelling H (1933) Analysis of a complex of statistical variables into principal components. J Educ Psychol 24:417

Janes KA, Yaffe MB (2006) Data-driven modelling of signal-transduction networks. Nat Rev Mol Cell Biol 7:820–828. https://doi.org/10.1038/nrm2041

Kim S et al (2016) PubChem substance and compound databases. Nucleic Acids Res 44:D1202–D1213. https://doi.org/10.1093/nar/gkv951

King TS, Lengerich R, Bai S (2016) What is the role of statistics in clinical research? The Pennsylvania State University. https://onlinecourses.science.psu.edu/stat509/node/2

Kumar R, Chow CC, Bartels JD, Clermont G, Vodovotz Y (2008) A mathematical simulation of the inflammatory response to anthrax infection. Shock 29:104–111

Law V et al (2014) DrugBank 4.0: shedding new light on drug metabolism. Nucleic Acids Res 42:D1091–D1097. https://doi.org/10.1093/nar/gkt1068

Leber A et al (2016) Modeling the role of lanthionine synthetase C-like 2 (LANCL2) in the modulation of immune responses to helicobacter pylori infection. PLoS One 11:e0167440. https://doi.org/10.1371/journal.pone.0167440

Leber A, Hontecillas R, Abedi V, Tubau-Juni N, Zoccoli-Rodriguez V, Stewart C, Bassaganya-Riera J (2017) Modeling new immunoregulatory therapeutics as antimicrobial alternatives for treating Clostridium difficile infection. Artif Intell Med 78:1–13. https://doi.org/10.1016/j.artmed.2017.05.003

Li N, Verdolini K, Clermont G, Mi Q, Rubinstein EN, Hebda PA, Vodovotz Y (2008) A patient-specific in silico model of inflammation and healing tested in acute vocal fold injury. PLoS One 3:e2789

Lu P, Abedi V, Mei Y, Hontecillas R, Hoops S, Carbo A, Bassaganya-Riera J (2015) Supervised learning methods in modeling of CD4+ T cell heterogeneity. BioData Min 8:27. https://doi.org/10.1186/s13040-015-0060-6

Lund K, Vase L, Petersen GL, Jensen TS, Finnerup NB (2014) Randomised controlled trials may underestimate drug effects: balanced placebo trial design. PLoS One 9:e84104

Machado D, Costa RS, Rocha M, Ferreira EC, Tidor B, Rocha I (2011) Modeling formalisms in systems biology. AMB Express 1:45. https://doi.org/10.1186/2191-0855-1-45

Meinert CL (2012) ClinicalTrials: design, conduct and analysis. Oxford University Press, Oxford

Newell EW, Sigal N, Bendall SC, Nolan GP, Davis MM (2012) Cytometry by time-of-flight shows combinatorial cytokine expression and virus-specific cell niches within a continuum of CD8+ T cell phenotypes. Immunity 36:142–152. https://doi.org/10.1016/j.immuni.2012.01.002

Nguyen DV, Rocke DM (2002) Tumor classification by partial least squares using microarray gene expression data. Bioinformatics 18:39–50

Phippard D (2015) Big data analytics: the next evolution in drug development. http://www.biprocessonline.com/doc/big-data-analytics-the-next-evolution-in-drug-development-0001

PhRMA (2015) Biopharmaceutical R&D: the process behind new medicines. http://www.phrma.org/report/biopharmaceutical-research-and-development-the-process-behind-new-medicines. Accessed 18 Oct 2017

Piantadosi S (2013) Clinical trials: a methodologic perspective. Wiley, New York

Porta M (2014) A dictionary of epidemiology. Oxford University Press, Oxford

Prilutsky D, Shneider E, Shefer A, Rogachev B, Lobel L, Last M, Marks RS (2011) Differentiation between viral and bacterial acute infections using chemiluminescent signatures of circulating phagocytes. Anal Chem 83:4258–4265. https://doi.org/10.1021/ac200596f

Principal component analysis (PCA) procedure. (2016) The Pennsylvania State University. https://onlinecourses.science.psu.edu/stat505/node/52

Ramsundar B, Kearnes S, Riley P, Webster D, Konerding D, Pande V (2015) Massively multitask networks for drug discovery arXiv preprint arXiv:150202072

Rigden DJ, Fernandez-Suarez XM, Galperin MY (2016) The 2016 database issue of Nucleic Acids Research and an updated molecular biology database collection. Nucleic Acids Res 44:D1–D6. https://doi.org/10.1093/nar/gkv1356

Romero K et al (2015) The future is now: model-based clinical trial design for Alzheimer's disease. Clin Pharmacol Ther 97:210–214

Schmidt BJ, Casey FP, Paterson T, Chan JR (2013a) Alternate virtual populations elucidate the type I interferon signature predictive of the response to rituximab in rheumatoid arthritis. BMC Bioinformatics 14:221. https://doi.org/10.1186/1471-2105-14-221

Schmidt BJ, Papin JA, Musante CJ (2013b) Mechanistic systems modeling to guide drug discovery and development. Drug Discov Today 18:116–127. https://doi.org/10.1016/j.drudis.2012.09.003

Segovia-Juarez JL, Ganguli S, Kirschner D (2004) Identifying control mechanisms of granuloma formation during M. tuberculosis infection using an agent-based model. J Theor Biol 231:357–376. https://doi.org/10.1016/j.jtbi.2004.06.031

Siettos CI, Russo L (2013) Mathematical modeling of infectious disease dynamics. Virulence 4:295–306. https://doi.org/10.4161/viru.24041

Sullivan L (2016) Hypothesis testing—analysis of variance (ANOVA). Boston University. http://sphweb.bumc.bu.edu/otlt/MPH-Modules/BS/BS704_HypothesisTesting-ANOVA/index.html

van Die MD, Bone KM, Burger HG, Teede HJ (2009) Are we drawing the right conclusions from randomised placebo-controlled trials? A post-hoc analysis of data from a randomised controlled trial. BMC Med Res Methodol 9:41

Viceconti M, Henney A, Morley-Fletcher E (2016) In silico clinical trials: how computer simulation will transform the biomedical industry. Int J Clin Trials 3:37–46

Wells BJ, Chagin KM, Nowacki AS, Kattan MW (2013) Strategies for handling missing data in electronic health record derived data. EGEMS 1:1035. https://doi.org/10.13063/2327-9214.1035

Wendelsdorf K, Bassaganya-Riera J, Hontecillas R, Eubank S (2010) Model of colonic inflammation: immune modulatory mechanisms in inflammatory bowel disease. J Theor Biol 264:1225–1239. https://doi.org/10.1016/j.jtbi.2010.03.027

White IR, Royston P, Wood AM (2011) Multiple imputation using chained equations: issues and guidance for practice. Stat Med 30:377–399. https://doi.org/10.1002/sim.4067

Wise A, Bar-Joseph Z (2014) SMARTS: reconstructing disease response networks from multiple individuals using time series gene expression data. Bioinformatics 31(8):1250–1257

Zhang L, Athale CA, Deisboeck TS (2007) Development of a three-dimensional multiscale agent-based tumor model: simulating gene-protein interaction profiles, cell phenotypes and multicellular patterns in brain cancer. J Theor Biol 244:96–107. https://doi.org/10.1016/j.jtbi.2006.06.034

Index

© Springer International Publishing AG, part of Springer Nature 2018
J. Bassaganya-Riera (ed.), *Accelerated Path to Cures*,
https://doi.org/10.1007/978-3-319-73238-1

Printed in the United States
By Bookmasters